THE HUMAN ORIGINS

Valentin Matcas, M.Ed.

Copyright © 2026 Valentin Matcas

All rights reserved.

ISBN: 9781973415138

DEDICATION

I dedicate this book to everyone eager to learn and develop continuously throughout life.

CONTENTS

1. The Quest for the Human Origins — 1
2. Comprehensive Human Existence — 52
3. The Human Intelligence and its Origins — 72
4. Scientific Speculations in Intelligent Disguise — 135
5. The Origins of the Human Civilization — 207
6. Bloodlines and Timelines, Origins and Development — 248
7. For Those Who Remember — 267

1 THE QUEST FOR THE HUMAN ORIGINS

There is more to the human origins, human development, human intelligence, and human civilization than the endless debate between Creationism and Evolution, because there is more to humanity than what ideologies and the current authorities want you to believe. Furthermore, when you study the human origins, you must reach beyond the moment when the first humans detached themselves from the firmament or from previous species, since there are other significant events in humanity's lifespan and achievement defining the entire timeline. While you must study everything, otherwise you risk understanding these important events only from simplistic empirical or ideological perspectives, ending up learning ignorantly what you already know, while following the crowd throughout unending debates. You want the entire accurate truth, because you already know the current simplistic theories, beliefs, speculations, and debates.

This is why when you study the human origins, you must find and you must understand everything about the origins of life, the nature and origins of this world, the nature of the human higher self and intelligence, the origins and debut of the human consciousness and human intelligent reasoning, along with all details related to the Creator of our world, to humanity, and to Life altogether. Additionally, it is important to know how all these affect you and how they affect your family, genetic line, and entire society, and how your family and genetic line originate, where and how it happened, under what circumstances, and with what status and privileges for you, for your family, and for the entire humankind.

This book studies the human origins along with the origins of life, human intelligence, human species, human development, human society, human civilization, and past civilizations, integrating humans, their origins, and their original and current condition in life and in the world in an elaborate comprehensive study.

Throughout this entire research of the human origins, meaning, and development, we study the actual answer to when humans and life originated on Earth or anywhere else, how it happened, where, for what purpose, who was involved, with what consequences, and how everything developed to become what we are today. We want to know everything, and more importantly, we want to know it in an accurate manner. While this is not only a simple desire to know everything in the world, but as we will see throughout the book, we must always know everything about the human origins in accurate details, because all these define directly the human meaning and humanity altogether, affecting directly each living human being, while influencing continuously their life, achievements, and development.

How can you ever find out the actual truth? The most common way is to study the beliefs of your own ideology, either religious, scientific, spiritual, national, cultural, political, or social, in order to find the truth. All ideologies are sets of beliefs, which means that all ideological knowledge is based on beliefs, while all beliefs might or might not be accurately true, but only believed to be true. Therefore, since not all beliefs are accurately true, while they are not even consistent from one ideology to another, it might not be enough, since you actually want to find the accurate truth. Beliefs can be true sometimes, yet how can you tell from the start which specific belief from which specific ideology is true, while everybody is made to believe in everything even unconditionally, regardless if it is true or not? However, if all current beliefs, ideologies, theories, and speculations are sufficient to you, then yes, you already know everything in the world including the human origins, through all your beliefs.

Another way to study the human origins is through all scientific information provided by science, and this is the easiest way. Yet you already know all science from school and from the media, which means that again, you already know everything in the world, and therefore you do not have to research anymore. This must be the actual knowledge about the human origins, exactly as it is presented by the current science in the theory of evolution and in the survival of the fittest. Species form other species in time, which form other species, which form human beings, this is how life originated and evolved, only on Earth, and this is who you now are. Yet if this is accurately true, then how does life develop? Randomly, through random genetic mutations, since this is the answer coming from the current science, yet they do not call it development, but evolution. Evolution itself means random development. Not continuously improved coordinated development, but only random development.

Is this actually the truth? Yes, it could still be possible, since everything changes and develops throughout time, throughout Life, and throughout the world, and therefore it could still be possible for entire species to form other species randomly by accident, that form other species randomly by accident, forming ultimately humans randomly by accident, who are, as already known, the most evolved and the most wonderful and extraordinary living beings in the universe, since this is what the current science states.

Yet as you already notice, this is correct only through the words *could still be possible* used in the paragraph above, making the above statement not entirely accurate, but only probably true, and therefore not accurately true. While we always seek to learn only the accurate truth. The current science uses words as 'still possible,' 'probable,' and 'theoretical,' while these never define accuracy, even though science always uses them to make you believe that they define accuracy or accurate truth. All theories are not accurately true, but only speculative. While we seek accuracy throughout this research, not probable possibilities, neither assumptions, nor theories, otherwise we could have embraced any ideology with ease, and we had all their answers already provided.

The phrase 'could still be possible' used in the above statement actually means that, even though many things change in time, this does not mean that humans have to evolve. It is true that human beings grow up and develop in time, as you do throughout your own life, physically and cognitively. Yet just because living human beings grow up and develop, this does not mean that entire species develop or evolve into other species just as well. Living human beings develop, while the entire human species develops on its own, through all living human beings composing it. However, with all living human beings developing within the entire human species, it does not mean that the human species had evolved from the monkey species as the current science states. Because with the monkey species still around, the entire theory of evolution promoted by the current science is false. The current science might add that only some monkeys and apes had evolved into human beings, while forming the primitive humans and then the normally developed humans, with the rest of the monkeys and apes never evolving into primitive humans and then into normally developed humans, and this is why all monkeys are still around.

There are even anthropological and archeological records proving this entire scientific statement, records that might be genuine or not, yet this is not anymore the theory of evolution. Because the theory of evolution states clearly that all species evolve one from another from another from another from another, while following exactly the taxonomy of organic life that is also provided in the theory of evolution, by Darwin himself. However, all scientists including Darwin prove the theory of evolution by stating something else, that only some living beings of a species evolve into the next species as they develop normally, but not the entire species.

Yes, the theory of evolution states clearly that species evolve one into another into another, while the current scientists state something else, that only some living beings of a particular species develop substantially into something else, an entirely new species. While in this manner, the current scientists actually answer a question with the correct answer of a different question, while actually proving the theory of evolution is false.

If the theory of evolution is not accurately true, it might compromise the entire taxonomy of organic life. However, taxonomy is only a classification of all species forming the organic form of life, while as all classifications, taxonomy itself is only algorithmic and therefore abstract, and can never define and explain the origins and development of life, because life itself is significantly more complex than all abstract algorithms, including classifications. Which means that taxonomy can only point abstractly to the origin of life, but it can neither explain it nor model its development in time.

Furthermore, even though taxonomy cannot explain and model the origins and development of organic life, taxonomy can always divide living beings and the organic life altogether into distinct species even accurately, yet it does so only in an abstract algorithmic manner, which always has its limitations, and it is important to identify them.

This is not a flaw in the theory of evolution, but it is only in the limitations offered by algorithms and by the entire algorithmic thinking. However, the theory of evolution should have never used abstract algorithms to model the development of the entire organic life, since it cannot match the complexity of life in the real world.

However, since the entire theory of evolution is only a theory, it means that the entire theory of evolution is only an assumption or a speculation, and therefore it can still be abstract and algorithmic, since it is only assumptive or speculative, never describing the actual origins and development of life as these actually took place. While we seek only accuracy, since we want to know exactly the human origins in life and in the real world, exactly as they took place.

Therefore, the taxonomy of organic life is true but only in an abstract manner, having its limitations. Therefore, expect the classification of organic life offered by taxonomy to be different than what happens in life and in the real world, because life and the real world are significantly more complex than any classification including the entire taxonomy, because all classifications are algorithmic and abstract. Similarly, computers and computer programs have their own major limitations compared to real life and the real world, because computers and computer programs use only algorithms, which are always incompatible with the living complexity of life and of the real world.

This is very important in our study, because it helps us know from the beginning that even if life is divided into kingdoms, classes, species, and living beings as taxonomy depicts, it is only an abstract empirical algorithmic classification of living beings and of life altogether, never depicting, following, or explaining the origins and development of life. Which means that, in real life, living beings are differently related to each other, in a more complex manner than everything presented by the taxonomy of organic life, because the abstract empirical algorithmic taxonomy is always incompatible with the entire complexity of life, and therefore it is always insufficient to model the origins and development of life.

However, in an abstract algorithmic manner, we can still state that living beings are divided into species, or that living beings form species. While in real life, and throughout our study of the human origins and development, expect an entire complexity of life, far superior to a simple classification.

Similarly, computers can never reach intuition and intelligence by using algorithms, because algorithms are always significantly inferior and therefore incompatible with intuition and intelligence, since both intuition and intelligence are main characteristics of life, they are part of life, while life itself is significantly more complex than abstract algorithms.

This is why the theory of evolution states clearly that species evolve one into another into another up to humans themselves, because it has to state everything according to the taxonomy of organic life, since the taxonomy of organic life is used ignorantly by the theory of evolution to prove itself. Because this is what Darwin came up with back then, and now this is what the current science must follow.

However, the current scientists state something else about the theory of evolution, that only some members of a species develop substantially into something else while coping with the environment, they develop into an entirely different species, and they do so in order to cope with the environment. While by coping continuously with the environment, nothing is random nor accidental throughout the development of life, but it is intuitive and intelligent, coordinated continuously in order to assure survival, subsistence, prosperity, and development.

Notice how the current scientists are more accurate compared with the theory of evolution itself, however, even the current scientists remain incapable to explain how living beings manage to develop from one species to another. Therefore, the current scientific explanation of the entire development of life is still the random accidental genetic mutations, because nobody can understand and explain intuition and intelligence as they are used continuously by all living beings to cope with the environment, and how entire large groups of living beings are capable to form together entire species, classes of life, and kingdoms of life throughout the entire development of life.

Therefore, if you want to know accurately the human origins, you must also understand accurately the human cognition. While if you want to know everything about the origins and development of the entire life, you must be able to consider the entire cognitive aspect of life, which might be harder than expected. This is why we include all human origins in this study, including the origins of life here on Earth and everywhere else, along with the origins and development of the human intuition and human intelligence, while doing the same for all living beings of all species, classes, kingdoms, and forms of life.

Why is the entire classification of organic life from taxonomy different than everything that we have in real life and in the real world? Is taxonomy itself false, making the entire theory of evolution false? Taxonomy itself is accurate, but only at its own empirical abstract algorithmic level, since all classifications are algorithmic, abstract, and empirical, remaining accurate but only at this low level.

All classifications are made by something, by a specific characteristic defining all items that you classify. In this manner, you can classify all the cars from the parking lot by top speed, obtaining a specific order or classification of all cars from your parking lot. However, if you classify your cars by size, you must make a different classification, since the order of your cars from the parking lot is different now by size. You can also classify bicycles by color, weight, price, and everything else that you find necessary. While you can even classify everybody from your office by age or by weight. While as you notice, you always classify all items in an algorithmic manner, by using if - then algorithms, in order to place everything in a specific order or classification.

However, as it is the case with cars and bicycle when we classify them by color, size, weight, or top speed, classifications can never explain how cars work, how people lose weight, or how the entire life develops from one species to another, but classifications only classify items, nothing else. The taxonomy of life cannot explain the origins and development of life, but it only classifies living beings empirically into races, species, classes of life, and kingdoms of life.

Throughout taxonomy, all living beings alive and extinct are classified empirically, by their physical appearance, and by all their physical characteristics, into all races and species. Furthermore, all species are classified into classes of life by similar empirical physical characteristics, and then all classes of life are classified into kingdoms of life similarly empirical, forming the entire taxonomy of life here on Earth. This is only the taxonomy of organic life here on Earth, yet even so, it is very accurate. Yet it is only empirically algorithmically accurate, incapable to address the entire complexity of life. Even so, the taxonomy of organic life is still helpful throughout our study of the human origins.

Therefore, even though taxonomy is empirically accurate, since all classifications are empirically accurate, expect to encounter a multitude of limitations. Because taxonomy, even though it is used by Darwin to explain and prove his entire theory of evolution, it cannot explain how species evolve, where the missing links are, how living beings are capable to cope with the environment while forming other species, how intuition and intelligence are used by all living beings to cope with the environment, and how intuition, intelligence, and the entire cognition work.

We already notice very specific errors of reasoning in the theory of evolution and in all scientific research using the theory of evolution. In the first place, you cannot use taxonomy, which is the classification of organic life, in order to prove the theory of evolution, since classifications cannot explain development itself, because classifications only classify items, nothing more.

Secondly, the entire classification of life from taxonomy is done only by the physical characteristics of all living beings, since all classifications are done by only one common characteristic. If you want to classify living beings by other characteristics, as cognitive, interactive, or existential, then you must form different classifications, different taxonomies. Therefore, you cannot use the theory of evolution and the current taxonomy to explain how living beings develop into entire new species while coping with the environment, because they always cope with the environment intuitively and intelligently, these are cognitive in nature, while the current taxonomy classifies the organic form of life by physical characteristics, not by cognitive characteristics. It is very important that you can identify this second error of reasoning throughout the research, otherwise you fails understanding life and its origins and development.

As a third error of reasoning, you cannot mix empirical algorithmic thinking with ideological consensual thinking, with intuitive thinking, and with intelligent reasoning, since some of these remain incompatible, compromising your results.

As a reference, theories are not accurate but only consensually valid, but only if they are validated by the current scientific consensus. Yet even if theories are validated by the scientific consensus, it does not guarantee their accuracy, because if they were accurate, then they were not called theories anymore, but accurate facts or even natural laws. As a reference, the current classical physics if full of laws and accurate facts, not of theories. While the current modern physics is full of theories, not too much of accurate facts.

Therefore, you cannot use theories throughout intuitive and intelligent research, because they are not accurate. You can still use theories, but only throughout consensual research, while obtaining further theories as results, also validated by the scientific consensus, yet they are not accurately true. Only intelligent research can guarantee accurate results, but only if you use accurate knowledge throughout your intelligent reasoning, never otherwise. However, if you ever use theories throughout intelligent research, your results are also theoretical, never accurate. The scientific consensus should also validate them, yet even then, your results are not accurate, but only consensually valid. This is why the current science is full of theories, none offering accurate knowledge.

As you notice, you must use only intelligent reasoning while studying life in order to obtain accurate results, because life itself is always accurate in the real world, as it causes all real accurate events from the real world. This is why you must always study life, humans, and humanity altogether by using only accurate knowledge throughout an entire intelligent research, never otherwise. While this is what we do throughout this book and book series "Human," in order to obtain only accurate results.

As stated, the current taxonomy classifies living beings by their physical characteristics into all races, species, classes of life, and kingdoms of life that you know well. However, all living beings develop from one species to another while coping with the environment by using all their abilities, including their physical, cognitive, and interactive ones. While the current taxonomy is made only by physical characteristics, which is insufficient, leading to discrepancies between the current taxonomy and the real world. Yet even so, the current taxonomy is the best that you can have from physical perspectives, while if you want to classify life from living, cognitive, or interactive perspectives, you have to make additional classifications. Because as stated, all classifications use only one characteristic at a time to classify items, never more.

Therefore, expect the real life to be different than the current taxonomy, because, even though minds and physical bodies are correspondent, they lead to different classifications of living beings in life and in the world. All differences are not major, but it is important to identify them, otherwise you fail, alongside the entire current science. Let us now see a major discrepancy between the current taxonomy made only by physical characteristics, and real life.

In real life, all living beings are classified by their living characteristics, not only by their physical characteristics or by their cognitive ones. Therefore, in real life, the entire development of life takes place in a different manner than everything presented abstractly by the current taxonomy. As you study the real life closely, you notice how all living beings live their life not within distinct species as the current taxonomy states, but all living beings live life throughout an entire succession of species, and it is important to identify it. More precisely, the life of all living beings is formed of an entire succession of species, as you can see it very easily in all living beings, with humans and frogs included. Because all frogs are born as fish, while frogs and fish are different species, and even different classes of life. While this is a major discrepancy between the current taxonomy based on physical characteristics and the real life.

Therefore, you do not have to find all the missing links between species as all scientists persist unsuccessfully, because these are never there, since species never evolve into newer species, but only living beings themselves develop normally throughout life from one species to another ever since the dawn of life. Species never evolve one into another, but only a group of living beings from a species develops substantially into something else, while forming together a new species.

All living beings grow up and develop, as you already know, since it is also your case. However, this does not mean that entire species of living beings evolve and transform into other species, as the theory of evolution states. All living beings develop on their own from the dawn of life, while forming throughout time alongside others an entire succession of species exactly as it is depicted by the taxonomy of organic life. Therefore, from a physical perspective, all kingdoms are formed of classes, which are formed of species, which are formed of races, which are formed of living beings, exactly as the current taxonomy depicts. With some or many discrepancies, caused by the incompatibility between the abstract classifications and the real life. While from an entire living perspective, you notice how the life of all living beings is lived throughout a succession of previous species occurring in the first stage of life. This is how frogs are born as fishes, transforming into frogs later on throughout life.

While you can see the truth about the origins and development of life yourself, since it is everywhere. All flies are born as worms, living life in this manner as worms at first, and then as flies. This is an accurate fact, since you can witness it firsthand. Flies and worms are different species, with worms preceding flies in the taxonomy of the organic form of life. Yet according to the current science, since all species evolve, the worm species evolved into the fly species, which is not correct.

Furthermore, the current science cannot explain the entire evolution of life, motivating it only on random accidental genetic mutations during transcription and translation. While by using a taxonomy based on living characteristics, or at least on cognitive characteristics, you can follow the development of life in the real world closely.

As seen, all living beings develop successively from one species to another, yet only in the first stage of life. This is also the case with humans, since the human fetus develops consistently from one species to another throughout gestation, until it reaches the human species in the last part of gestation. Notice how all living beings live life in the same succession of species as it had been the case throughout time with their entire species ever since the dawn of life. This is a major coincidence, leading to the accurate understanding of the origins and development of all living beings of the organic life with humans included, which is different than the accidental genetic mutations promoted by the current science.

The current science with its theory of evolution is wrong, because we do not witness the entire worm species evolving into the entire fly species, but only individual living beings as they develop normally throughout life, being born as worms, and then developing gradually into flies as they grow up, in a matter of hours or days. This is normal development or normal growth, and we can witness it ourselves, which means that it is accurately true. However, the entire worm species does not evolve into the entire fly species, but only individual living beings grow up or develop throughout life from one species to another.

Can we still state in an accurate manner that the entire worm species had evolved into the entire fly species? We already have two different topics, and we must distinguish them accordingly. First we have the individual living beings born as worm and then growing into flies throughout life as you notice it yourself, and therefore this is always accurately true. While the second topic is the entire fly species presumably evolving from the entire worm species as the current science states, which might or might not be true, since it is only a scientific theory, the theory of evolution.

While we must always be very careful not to answer a question with the true answer of a different question, since this is very common throughout ideologies, including scientific ideologies as the current science. Because we can simply state here that since all flies in the world are born as worms, which we can witness ourselves and therefore it is accurately true, it means that the entire fly species had evolved in time from the worm species, which is not true, because we answer a question with the correct answer of a different question.

We cannot state accurately that the entire species of flies had evolved from the entire species of worms. However, we can always assume that since all flies are born as worms, then the entire species of flies evolved from worms, exactly as the current science states. Yet this is only an assumption, a theory, a speculation, or a supposition, which might or might not be true, and you never know. Furthermore, since it is not an accurate fact but only a theory, we cannot use it throughout our intelligent reasoning during our study of the human origins, because we accept only accurate facts in order not to compromise our results. However, we can use our first discovery as an accurate fact, that all flies are born as worms and they grow up or they develop throughout life into flies, because we can witness it ourselves.

The terms growth, development, and evolution are relatively synonymous. However we will use more the term development, as we keep it consistent to everything that we witness in real life, in order to maintain our research accurate. The term evolution is more abstract, matching theories and abstract classifications of life, while the term development matches life itself at all its levels, since development is a main characteristic of life.

As a reference, the human development is of the third intelligent level, since humans are both alive and intelligent. While you must always maintain compatibility in your reasoning and entire research, in order to assure accurate results.

The current science is full of theories, while by definition, all theories are not accurately true. Accurate knowledge, accurate facts, accurate concepts, true events, mathematics, and natural laws are always accurately true. However, theories and many believes are accurately true only when they are proven right. Until then, they might or might not be true, and you cannot use them throughout your intelligent reasoning, since they might be wrong, compromising your results. Similarly, the entire theory of evolution promoted by the current science is not accurately true, but only an assumption or a theory. We cannot use it throughout our study of the human origins as an accurate truth, because it is only a theory or assumption.

Throughout our study of the human origins, we will use only intelligent lines of reasoning, in order to obtain only accurate results guaranteed. All intelligent lines of reasoning contain only accurate knowledge, in order not to compromise results. Therefore, we cannot use theories and beliefs throughout our intelligent lines of reasoning, unless we can prove them accurately true.

We already notice the distinction between debates and intelligent lines of reasoning, or between ideological beliefs and intelligent lines of reasoning. However, as already stated, if we manage to prove accurately that various scientific theories are accurately true, or that various ideological beliefs are accurately true, we can use them throughout our intelligent lines of reasoning, because only accurate knowledge remains compatible with intelligent lines of reasoning.

We also notice that the current science uses theories as accurate knowledge throughout its research, compromising all current scientific research. However, as long as the current science calls all its results theories, they are still true, but only in theory. Because if you use theories in your research, you end up with more theories. While we cannot do so, because as always stated, we seek only accurate truth in our research of the human origins, and therefore we cannot use theories or beliefs if we cannot prove them accurately true.

It is the same with frogs, since they are born as fishes. This is your missing link, since these particular frog fishes even grow legs in their first days of life, as you can even witness the legs growing. Which is an accurate empiric fact, since you can observe it yourself, firsthand. However, you cannot state accurately that the entire fish species had evolved into the entire frog species, mostly since all fish species are still present in the world.

Flies and the great majority of insects start their individual life as worms, which is an accurate fact. Frogs start their individual life as fishes, which is also an accurate fact that you observe yourself. At a closer study, you notice how this is the case with all living beings, since all living beings start their very early life as other species, while this is important to consider.

All living beings start life as their own previous species, morphing throughout them, individually, in a matter of days, weeks, or months, to become what we have today. You can also study closely the human fetus as it develops gradually resembling all previous species exactly as these are depicted in taxonomy. You notice how forms of life are not divided into species of life, but forms of life are divided directly into individual living beings, while the current science ignores it or hides it.

As you study the theory of evolution closely, you notice how it answers a question with the true answer of a different question. The theory of evolution uses the entire taxonomy of the organic form of life to prove that species evolve, which is an improper line of reasoning always generating erroneous results, since it simply answers a question with the true answer of another question. Because the theory of evolution includes two subjects: the taxonomy of the organic form of life which is true, and the evolution of the organic form of life from one species to another which might or might not be true, but you cannot confuse the two. However, as long as the current science calls the theory of evolution a theory, it is only theoretically or speculatively true, but not accurately true.

Furthermore, you notice how the theory of evolution never

addresses the actual development of the organic life from one species to another, but only the evolution of life, which is different. Because the term 'evolution' is abstract, matching taxonomy itself with its entire abstract classification of life. While the term 'development' is an actual conception, which is natural and alive, capable to depict life directly, as it is even a major characteristic of life.

Notice the continuous consistency of the current science as it uses only an abstract scientific consensus to validate itself but not accuracy itself as it should, always providing mostly theories, which are abstract, but not accurate knowledge as it should, while always using specific abstract concepts as 'evolution' while addressing life and the real world, but not natural conceptions as 'development' as it should.

Is this the actual missing link, the frog itself, when it was little, since it had legs? No, because all individual living beings develop in this specific manner throughout life, from one species to another, naturally, while species evolve only abstractly into other species, but only in the theory of evolution and in the current abstract taxonomy of life, not in real life. Life itself does not develop naturally from one species to another, but only the living beings composing species do, as it is the case with the frog growing legs when it is little, when it is only a fish. Because in real life, species themselves are part of living beings, not the other way around. While in an abstract taxonomical manner, living beings form species, not the other way around.

As stated, it is an error of reasoning to mix abstract concepts as the entire taxonomy of life, with living intelligent conceptions as the development of life. While, if we are very careful, we manage to avoid all errors of reasoning. With the current science failing continuously, while remaining incapable to distinguish between the classification of life, the development of life, and life altogether.

This is why the current science calls it only evolution of life, but not growth of life or development of life, because evolution is abstract, matching closely the current abstract taxonomy of life. While in this manner, the theory of evolution addresses only the taxonomy of life, but never the real life from the real world as it originates and as it develops from the dawn of life to everything that we currently have, the entire diversity of life. While throughout our study, we seek to follow the real natural life, as it originates and develops, with the human life, human interconnectivity, and human cognition included. While if we are very careful, and if we manage to use only accurate knowledge throughout intelligent lines of reasoning, we manage to model and understand all human origins along with the entire human development.

You cannot see the entire human development from one species to another in the human fetus, since humans spend the first stage of life in the womb, and are harder to see. However, as you study the human fetus closely in pictures, you see it developing through various species, right there in the womb. It is more visible with all species below reptiles, as it is the case with frogs growing legs when they are young, when they are fish. While the chicken is even a genuine living pterodactyl inside the egg at a specific stage of life, but only inside the egg, because chickens are also born as chickens when they hatch, not as reptiles or as amphibians before.

There are many missing links in each living being, encountered mostly throughout gestation. Because new species cannot give birth to new individuals, but only new individual living beings can form new species. It is the other way around, with living beings going through various species in the first stage of life, while the current science ignores it. Therefore, species do not evolve, but individual living beings develop directly, throughout life, and more importantly, individual living beings develop from one species to another throughout their own individual life.

Note how science states the contrary, that species evolve from one another, and this is why the individuals composing these species are different from one species to another. While as stated, it is the other way around, individual living beings develop throughout life from one species to another since individual living beings are of many successive species in their first stage of life, determining in this manner the specific species that they are later on in life. It is a difference in statements and concepts, as we already notice.

While we must still find out why there is a coincidence between the succession of species throughout time as depicted by the current taxonomy, and the succession of species that all living beings undergo throughout the first stage of life in order to become the species that they actually are. Because as you study the frogs or the chickens inside their eggs, you notice them going through a specific succession of species, which is exactly the succession of species presented by taxonomy.

This observation helps us throughout our research of the human origins. In the real living world, the existential arrow points towards individuals and not towards the species that they form. This is an important detail in our study of the human origins, implying that the essence of life is within living beings but not outside. Which means that the individual living beings give life and meaning to their entire species and then to the entire form of life.

Do humans come from apes? Are apes actually the human origins, since this is what science states? Since flies come from worms, and since frogs come from fish, it might seem proper to state that humans come from apes. Yet this is false, because you end up answering a question with the correct answers of two other questions. Because humans are not born as apes, in order for humans to be apes for the first year of life, the way flies are born as worms and frogs are born as fishes.

Yet there is more to consider, since humans tend to resemble little monkeys during the last stages of gestation, they even have a tail, and if you are ever able to prove that that is a monkey, then your answer is positive. Whatever that is in the last stages of the human gestation, a monkey, an amphibian, a cat, or a reptile, it proves that it is not the human species as a whole evolving, but each individual living human being develops in this specific manner inside the womb, as a distinct living being. This happening to all living beings, including humans.

This means that humans do not originate from monkeys, but humans transition throughout all previous species of life during gestation, with the last species as any mammal species, whatever you manage to distinguish in the pictures.

The human fetus maintains consistency with the development of the entire humanity from one species to another. As you study the human fetus closely, you notice how humanity spent a long time as a particular past amphibious species, which resembles more with the common grey than to any amphibious species currently present on Earth. While as a last species, the human fetus resembles more to a feline or to a canine than to a monkey, which means that if humanity did pass through the monkey species, it did so very shortly. Currently, some scientists consider humans as part of the ape species, while others consider the human species independent from the monkey species.

At a closer study of taxonomy, you might notice various species converging into one unique species. While as you study humans and the human fetus closely, you notice how some humans went through the feline species, while others went through the canine species, merging in any manner later on as a unique, compatible human species.

While again, we must figure out how we have this persistent coincidence between the particular succession of species developing into the human species throughout billions of years, and the fast succession of species undergone by the human fetus in its first stage of life in only months. The answer is in the current taxonomy based on physical appearances incapable to model an accurate development of all living beings throughout species, and in the current definition of a living being, since as you notice, the little white worm is not considered by the current science a living fly being, the chicken egg is not considered by the current science a living chicken being, and the human fetus is not considered by the current science a living human being. Yet as a living being, it does not matter if you are born directly as a cell, in an egg, or in a womb, you are always a living being, starting with the moment of conception. Because if we had an entire taxonomy based on living characteristics or even on cognitive characteristics, then we already had our answer, while all living beings were considered by the current science alive starting with conception.

As you study closely the entire human gestation, you notice how living human beings start life as a single cellular living being, going through a multitude of sequential species exactly as these are classified in the current taxonomy, becoming of the human species later on during gestation. Which means that all humans, as an organic form of life, have their origins in a cell, but not exactly in a monkey or ape. While this is the case with all living beings, since all originate in a cell at the moment of conception. It is the same with all living beings, while coincidentally, it is relatively the same cell for all living being. As another coincidence, this is the actual original cell that existed back then at the dawn of organic life, making all future species possible.

While as already stated, we must find out why we encounter so many coincidences. So far, we notice how we must understand the development of life from living perspectives, not only from physical perspectives as the current theory of evolution seeks unsuccessfully. Yet if we cannot, we must understand the development of life at least from cognitive perspectives, which might be easier to achieve. However, we must seek to learn at first everything from physical perspectives, since it is easier, while assuring implicitly an empiric perspective, which is accurate by default. Otherwise, throughout living and cognitive perspectives, we end up with speculations and assumptions as results, since cognitive details are harder or impossible to see firsthand.

However, it is already surprising for the current science to keep the theory of evolution in use for many centuries, through a continuous consensus coming from billions of teachers and scientists, while it is obviously obsolete, standing now in our way as we learn everything about the human origins, human development, and human meaning. Why?

Because the theory of evolution replaced successfully Creationism in the West, while helping overturn an entire world order in the process, based on a religious ideology that you know well. While currently, the entire authority in the West is based on legal ideologies called jurisdictions, on political ideologies that you know well, on all social ideologies reaching you, and on a particular scientific ideology that you might ignore, which is the current science.

Is science an ideology? No, since science should always be based directly on the natural laws of this world, always in an accurate manner. However, the current science is based on a scientific consensus instead, validating all scientific truth consensually, while making all theories valid, but only consensually valid, and therefore not necessarily accurate. By definition, all ideologies are systems of beliefs, and therefore ideologies are always consensual. While all scientific theories and assumptions are only consensually valid, similar to all beliefs, not necessarily accurate. This makes the current science a genuine ideology, a genuine scientific ideology, similar in structure to all ideologies, all political, social, and religious ideologies that you know well. With the current authorities controlling all these, while using them to control everybody.

The current science is based on a scientific consensus that validates all its theories and scientific beliefs, making it a genuine ideology, a scientific ideology. However, since the current science has priority in court, it stands above all political ideologies, social ideologies, and religious ideologies, at least in the West. While in the East, there are mostly political ideologies in control, or only religious ideologies, or even the same scientific ideology that you find in the West. While through its consensual theories as the theory of evolution and the big bang theory, the current science decides the human meaning in life and in the world, diverting everybody from the actual human meaning in life and in the world.

Because as you will see throughout the book, once we are successful to find the origins of life, of humanity, and of all living human beings from all perspectives, we manage implicitly to find the human development and the human meaning, exactly as they come from Life. While until then, since we are still in the beginning of our study, we notice all these conspicuous coincidences, discrepancies, and strange hidden and revised terms and concepts, which actually help us in our study.

The current science uses the concept 'organic life' for all life, even for cells, in a conspicuous manner. While the current science might do so deliberately or only through ignorance. However, through its etymology, organic life means life lived in form of organisms. This means that sometimes in the past, all cells were considered alive, and there were two concepts in use to classify life, the cellular form of life, and the organic form of life. However, for various reasons, the concept 'cellular form of life' or 'cellular life' was erased from the current science, leaving behind only the concept 'organic form of life,' or 'organic life,' used now to depict both the cellular form of life and the organic form of life.

Throughout this entire book series "Human," we notice a multitude of circumstances when the current human society and the past ones revise the human knowledge in every manner while fulfilling various agendas, as it is the case now with life and organic life meaning the same, while only etymology uncovers the truth.

Therefore, all multicellular organisms are of the organic form of life, never unicellular organisms. Yet the current science persists to classify erroneously all prokaryotes as organic life, while they are single cellular organisms, and therefore part of the cellular form of life. Furthermore, the current science considers only prokaryotic cells as living beings, never eukaryotes, because eukaryotes form organisms, while according to the current science, eukaryotes cannot be alive, since only the entire organism is alive.

We notice a coincidence in statements when the current science considers only prokaryotes of the organic form of life, never eukaryotes, while considering only prokaryotes alive, never eukaryotes. While in this manner, the current science considers only the organic form of life alive, while hiding all the forms of life found below the organic form of life, in order to hide the origins and development of life. Consequently, according to the current science, all humans start their life at birth, with the human fetus never considered alive. Because if the human fetus is considered alive, then all abortions are human murder, and must be considered accordingly by the entire system of justice.

As you study all organisms closely, you notice how the organic form of life stands right on top of the cellular form of life, having it at its base continuously, with all cells fulfilling continuously by specialization the entire organism. It is the same with all cellular components while they fulfill continuously the entire cell, further on fulfilling the entire organism, and then the entire society alongside everybody else. While you also notice how living beings are not only of the organic form of life, as it is the case with all organisms, plants and animals, but also of the cellular form of life, as it is the case with all prokaryotic and eukaryotic cells.

All cells are living beings of the cellular form of life, while all organisms are living beings of the organic form of life. However, as you already notice, you understand this statement only from physical perspectives, only as the current science defines living beings, and only as the current science maintains the current taxonomy of life based only on physical characteristics, making this entire statement questionable and debatable. However, once you switch to living and cognitive perspectives, you understand life exactly as it is. Surprisingly, in the past, people could understand life exactly as it is, while the current society works hard to revise some or many words, terms, concepts, and entire accurate knowledge used in the past. It is called revisionism, and since it ends in 'ism,' it is a social ideology, yet very similar in structure to all ideologies ending in 'ism' that you know well. When exactly will we be able to switch to living and cognitive perspectives in order to understand all these?

Until then, you notice how from physical perspectives, all living beings of the organic form of life have their origins in a cell, the same cell found at the moment of conception at the birth of all living beings, and the same cell that was at the origin of all organic life billions of years ago. Which is coincidental. Because the organic form of life stands right on top of the cellular form of life, with this particular cell at its base, but only in the very beginning. Therefore, humans do not have their origins in monkeys or apes, but in a very specific cell, the ovule, which is the human cell during conception marking the birth of the living human being, but which is coincidentally similar to the cell found at the origin of the entire organic form of life.

What exactly can it be so different and so important from cognitive perspectives, in order to help us understand clearly all these? The entire cognitive aspect of life, which is the entire cognitive perspective. While the physical perspective of life is similarly important, only that you must always have the two, the physical and the cognitive, in order to understand life. While in order to be able to follow an entire cognitive perspective of life, you must understand all the basic cognitive concepts first, while these might or might not be in the current science and therefore in the current human knowledge. Yet if they are not, you must uncover them from the hiding as they are currently revised, or you must discover them altogether, which might be more complex than you assume. Until then, we continue with our physical perspective, until we manage to switch to living and cognitive perspectives.

Therefore, humans do not come from apes, cats, and horses, but all individual living human beings develop throughout the entire organic life of Earth during gestation, being part of Life, part of Earth, and part of the organic form of life in this manner, entirely. While this is a living perspective. Humans do so as individual living human beings, which is an extraordinary achievement for Life. Because as you study Life herself along with all living human beings closely, you find all living human beings with you included sharing the same origin with Life herself. While this is the living human origin, part of the human origin.

Do people come from monkeys? No, but humans and all living beings have a common origin within Life herself, as they form and maintain Life herself. This is our answer so far of the living human origin, while as you notice, it is not a theory, assumption, or speculation, but an accurate fact, it is accurate knowledge.

Yet there is more to consider, since humans and all living beings are not only their physical bodies, but they are also their mind, cognition, or cognitive system just as well, along with much more. You are always mind, body, and soul, as one. This is why humans have a multitude of origins, since humans are more than the physical human body. Therefore, you must study the human origins and the human development from all perspectives: living, empirical, physical, cognitive, objective, spiritual, biological, historical, developmental, and social. Which means that you have to undergo a comprehensive study of the human origins, and while doing so, everything must remain consistent from one perspective to another, otherwise this entire study of the human origins fails. While as you notice, we are already further in our research and mental model of the human origins than what the current science and most of the ideologies of Earth have to offer.

All living beings of the organic form of life have a common origin in a primordial cell, not only humans, while it is the same cell. However, all living beings of all species transition from one species to another differently, while passing only through very specific species, but not through all. As a reference, humans might pass through felines or canines, while determining them to love cats or dogs throughout life, however, humans did not pass through chickens or through other birds, because as reptiles, they transitioned directly to mammals. While other reptiles transitioned to birds, making all birds possible, with all chickens included. More precisely, all the flying reptiles transitioned to birds or went extinct, others transitioned to mammals or went extinct, while the rest remained reptiles or went extinct. Currently we know only of birds and mammals, yet reptiles might have transitioned to other classes of life during that major cataclysm, yet they went extinct.

Because in general, throughout major cataclysms, we notice the formation of significant species and classes of life. Yet how exactly do living beings know how to make new species and classes of life? They use their minds to do so, exactly as you use your mind in order to fulfill all your needs, which is a very tedious and very demanding process. While as long as we cannot switch to cognitive perspectives, we cannot explain how living beings use their minds in order to live their life and in order to make new species throughout significant changes of the environment. Yet we can still state that species match their environment correspondently, while correspondence itself is a natural law of the universe and a major characteristic of life, making it possible. While when environments change significantly, the living beings forming entire species must change their bodies and minds significantly in order to be able to cope with the environment.

Again, if we knew more about minds, thinking, and cognition in general, we could elaborate this subject more, and we did so by switching entirely to a cognitive perspective. Yet we still know that brains are minds from a cognitive perspective, while only from a physical perspective they are seen as brains. While all vertebrates have brains and therefore all vertebrates think. However, even invertebrates have brains, of a different type, using them to think. All living beings think, while we still do not know how living being think. However, you always think, and therefore, once you understand how you think, you know how all living beings think, since it is the same. The current science explains cognition with the continuous change in the action potential throughout all neuronal connections within the brain, since they can always detect it by using electroencephalographs, yet the current science does not know how everything takes place, while making only assumptions and theories, nothing accurate. While once you cannot understand cognition itself, you cannot understand anything, including how living beings know how to change their own appearance significantly while forming entire new species.

We are stuck alongside the entire current science, only that now we know what to look for, since we must learn how to switch perspectives entirely from physical to cognitive in order to understand everything. While as you notice, even the current science does not know how to do so, because the current science attempts to explain all cognitive processes from physical perspectives, through neurons, neuronal networks, and action potential, which are always physical.

Study the fetus inside the chicken egg, to see it resembling to flying reptiles towards the end of the chicken gestation. That prehistoric animal is alive, right there inside the chicken egg, which is the case for all the chickens in the world. While it is not too hard to stop the chicken gestation right then, in order to make the living pterodactyl species possible once again.

Which could mean that long, long ago, chickens looked in that specific manner. While the existential arrow points towards the individual chicken, and not towards the entire chicken species, since it is actually the living being developing in this manner, and not the distinct living species of the organic form of life. The chicken species does not evolve from reptiles, but the individual living chicken lives life as a reptile very shortly during gestation, probably signifying that all chickens used to be reptiles some time long ago, after they were other species, as frogs and fishes. Yet there is more to consider, and we will see it throughout the book.

How does life develop from one species to another, and how do living beings grow in the first stage of life from one species to another, yet always the same succession of species? The two are connected and it cannot be otherwise, while the only answer that we have is that species do not know otherwise, but only how to follow this specific succession of species, nothing else.

There are more perspectives to follow besides the physical

and cognitive perspectives, as the living and social perspectives. While by following them, we manage to uncover all human origins, one through each perspective, as the origin of the human intelligence, and the origin of the human civilization.

Therefore, from social perspectives, we consider a multitude of living human beings forming communities, societies, and entire civilizations, always in a specialized manner. As you study the current human society and civilization, you notice how living human beings never behave randomly, but they work only within very distinct domains, as in the food production, food supply, shelter, education, transportation, security, clothing, medicine, and communication. These domains are found in all species and in all forms of life, because all animals of all species fulfill these for themselves and for everybody else. Furthermore, as you study organisms, you notice how all cells are similarly specialized while fulfilling similarly specialized needs for themselves and for the entire organism, making all cells genuine living beings, while all cells make entire organisms correspondently similar to entire societies and civilizations. Furthermore, as you study cells, you notice how all cellular components are similarly specialized in these distinct specializations found everywhere, as food, shelter, development, recover, and transportation, in all forms of life, classes of life, species, living beings, and in all packs, communities, societies, and civilizations, again in a coincidental manner.

While it is even easier to understand cells as entire civilizations of living beings than as individual living beings, with all specialized cellular components working continuously towards fulfilling all cellular tasks by specialization. Yet this is also the case with entire organisms, since all organisms are entire civilizations of specialized cells that know well how to perform their own specializations in order to make the entire organism possible, while these specializations are always related to the same specializations found in the human society: eating, recovery, circulation, ventilation, communication, and development or education.

Something transitions from within cells at the organism level and then outside in society maintaining consistency, and we must know what. While we cannot see it from physical perspectives too well, but only from social and living perspectives, and hopefully from cognitive perspectives, once we learn how. Yet if we want to switch to cognitive perspectives, we must know everything about cognition, including how living beings think and how humans reason.

It seems that organic life cannot even develop from one species to another, because species cannot give birth directly to their individuals composing them. All individuals composing species have to develop themselves during gestation, and they do so only throughout the entire succession of species that they have been. This means that the organic form of life is formed naturally and directly of individual living beings, not of species. However, you can classify easily all living beings into species, not in real life since this is not the case in real life, but only in an abstract manner, using taxonomy.

All living beings of the organic form of life are relatively similar, while they have to follow the same pathway throughout their development as all their ancestors did, developing now only in one lifetime from that original egg standing at the base of the organic form of life, to become in a matter of weeks or months, during gestation, what their previous generation had been, and then to take it from there and continue developing throughout life as the actual living beings of their current generation, in any manner it is more suitable for the specific environment where they find themselves, and for the specific meaning that they fulfill.

While now we observe distinct cognitive development and physical development taking place continuously, because as stated, living beings are bodies and mind as one. Which means that you cannot study the entire development of the physical body throughout gestation, throughout life, throughout the ages of Earth, and throughout the organic form of life, if you do not consider the origins and development of minds and cognition simultaneously. Because all living beings are bodies and mind as one, with some being significantly more than this, as it is the case with humans.

Since all living beings develop in this specific manner, having only one cell as origin, this means that the existential arrow points not only towards each individual living being and within each individual living being, but it points to these cells composing them directly, and then to all cellular components. While in this manner, the actual living beings are within cells, while the entire cell is their civilization, which is similar to humans and human civilizations, while always following the same specializations at all levels of life. More precisely, the existential arrow points inside each individual living being composing species, and not outside towards the entire species, towards the entire ecosystem, towards the entire reality. Therefore, whatever it is at the origin of everything, it is still found within each individual, and not outside.

This existential detail is highly relevant while showing its presence here, since it makes its presence throughout all models of this book series "Human." The truth and the essence of everything is within, not only in the outside. It is an inner, living, cognitive nature found alongside the objective, physical nature. The meaning of life is cognitive, physical, living, social, and existential. However, since the cognitive is found within or at the base of the physical, the meaning or essence of life is within.

Notice how there are already a multitude of ideological beliefs stating the same, while other ideological beliefs state the contrary along with everything else, so how could we know from the beginning which one is true? This is why, in order to study the human origins, we do not have to study only some extraordinary event from the past when humanity first walked the Earth, because this origin is found within all human beings even today. Even you have this reminiscence of the actual human origin within you, while this origin coincides not only with the origin of the humankind, but it coincides with the origin of the organic form of life, along with the origin of the human mind and human intelligences, along with the origin of the human development and human existence, along with the origin of the human civilization which is similar for all previous civilizations of Earth, along with the origin of the current human civilization, and along with much more, since you have it within, right now, the origin of life itself.

You can feel it just as well, since your origins stand at the base of all your needs, feelings, and meanings in life and in the world. You have Life in you, and you must always consider it, since it always defines your own meaning in life, just as it had defined the meaning in life for all species and intelligences that you had been before, counting in large numbers, with all their intelligences still alive today, in you, while forming you altogether.

This is why we have to know everything about the human origins, how everything took place, why fetuses resemble other species throughout gestation, how they come to be in this manner, how exactly are humans born, through what specific processes, how does organic life originate, what is organic life, what is life altogether, what are the origins of the human mind and of the human intelligence, how do these develop, and why. Even more, you do not want to consider only sets of separate, distinct answers generated by this research of the human origins, but you want all your answers to correlate, interconnect, and remain consistent, not only among themselves, but to remain consistent with all accurate facts of this world, the entire human knowledge.

More precisely, you want an entire study of the human origins and development, but not an enumeration of individual modular information. Because only through consistent, interconnected accurate knowledge, you are certain that everything is true, accurate, and therefore undebatable. While we already have their major correlation right from the start, since everything originates in one point and in one moment in time, for the same reason, matching in this manner all lines and lifelines of causality. This takes place at the moment of conception in your own case, while it had also happened at exactly the same time and place and for the same reason for the entire humankind and human civilization but earlier in the distant past within this world, and then it happened elsewhere in the wider world further in the distant past.

In order to achieve all these, we have to undergo a very specific, detailed, comprehensive study, which is actually an intelligent mental model that forms in your own mind the intelligent conception of the human origins. I refer to it as an intelligent mental model, since it is found in your own mind right now as you reason intelligently alongside this book. This is the introductory stage of your intelligent mental model for the human origins. It is you making this intelligent mental model, in your mind, as you reason intelligently in parallel with this book, since this is how the human cognition functions. Just make sure that your reasoning remains free of irrelevant beliefs and entire ideologies, in order to maintain it accurate.

All living beings develop on their own even throughout gestation, and they do so through their own genetic material. More precisely, humans have to develop from scratch throughout gestation, directly from the unicellular life that took place billions of years ago, throughout gestation alone, because this is the only thing that they know how to do.

How do living beings know how to do everything that they do? All living beings fulfill needs, while everything that all living beings do is meant to fulfill needs. It is the same with you, since everything that you do you do in order to fulfill needs, while you do everything in a specialized manner. However, everything that you do originates in your cognition, in all your cognitive abilities, which are similarly specialized while matching all these specializations found within cells and organisms, and in the outside world in society and in the human civilization. We also notice how you must use your intelligent reasoning continuously while you fulfill all your needs for the entire organism in the outside world, while currently, by reading this book, you fulfill your need for intelligent learning, which is part of your intelligent human development. But how exactly do you know how to reason intelligently while developing yourself during gestation? You know it subconsciously, through various subconscious cognitive abilities. Because as a living human being, you have physical abilities as running, hearing, and breathing, you have cognitive abilities as learning, remembering, feeling, and reading, along with many social and spiritual abilities.

Your cognitive abilities relate to your mind, while these are conscious abilities and subconscious abilities. However, all human abilities overlap, since you must use your mind throughout all your physical tasks, including walking, writing, and jumping. Yet in general, all physical abilities are made possible through the human body, mostly by using the muscular system, while all cognitive abilities are made possible by using the human mind.

The human mind is formed of two main intelligences, the main conscious intelligence, and the main subconscious intelligence. These main intelligences contain all your cognitive abilities, the conscious and subconscious ones. You know well your conscious abilities, as your reading, learning, counting, and memorizing conscious abilities, along with all your subconscious abilities, as your digestive, recovery, reproductive, and developmental abilities. However, when you study these abilities closely at their own cognitive levels within the human mind, you find them as inner intelligences, very similar in structure with your conscious intelligence and subconscious intelligence.

You are always a living human being in the outside world, you are a specialized living human being in the human society, while within the human cognition, you are only the conscious intelligence, because you cannot access and therefore you cannot control your subconscious. while as you notice, within your cognition, you are a normal, distinct, living intelligence. All your subconscious abilities are intelligences, they are the multitude of subconscious intelligences forming together the subconscious mind by specializations, as your digestive subconscious intelligence, recovery subconscious intelligence, developmental subconscious intelligence, and reproductive subconscious intelligence, while as you notice, some of them have their own organs and entire bodily systems. More precisely, from a physical perspective, these are your organs and bodily systems, while from cognitive perspectives, they are your subconscious intelligences.

Your eating or digestive subconscious intelligence has its own bodily system, the entire digestive system, yet it is very similar to you the conscious intelligence, since it knows well everything about food, eating, and digestion, and it is very pertinent in everything that it does. While for everything that it cannot do itself, it sends needs and feelings to other intelligences to help out, as it sends needs to you the conscious intelligence to find food in the outside world, or as it sends needs to the circulatory subconscious intelligence to transport all nutrients to all cells, or to the excretory subconscious intelligence to take out all the waste, or to the recovery intelligence to cure it when it gets sick.

All your subconscious intelligences send each other needs while they fulfill all their specialized tasks within the organism, making the entire organism possible. In this manner, they form entire systems of intelligences working together in a harmonious manner, while making the entire organism possible from all cognitive perspectives.

Your subconscious intelligences even send you needs and feelings, in order to help them in the outside world, since as a conscious intelligence, you are specialized in the outside world. While as you notice, we maintain our cognitive perspective. Because from physical perspectives in the real outside world you are the physical body, while from cognitive perspectives within your mind or cognition you are the conscious intelligence. While as a conscious intelligence, you are specialized in the fulfillment of all needs in the outside world, or you are specialized with the coordination of the entire organism as you fulfill your needs in the outside world.

How do all intelligences know how to do everything that they do in their own specialized manner? Similar to you the conscious intelligence, all intelligences have a multitude of abilities helping them do everything that they must do. While at a closer study, from a cognitive perspective, all abilities are specialized inner intelligences themselves, because all intelligences are systems of intelligences themselves, all having their roots within cells, cellular components, and further below.

All living beings are body and mind as one, while all minds are formed of intelligences in very large numbers, one for each conscious and subconscious ability, which are very numerous. All cells and all cellular components are living beings, as they are all mind and body as one. Cellular components are mostly proteins and enzymes, while they are all body and mind, with all minds formed of a multitude of subcellular intelligences, together forming in very large numbers all the necessary specialized systems of intelligences tending to the entire cell.

As you notice, everything that you do throughout life, you do in order to fulfill all your intelligences from a cognitive perspective. However, from physical perspectives, you do everything in order to tend to all your cells. While from social perspectives, everything that you do in life, you do in order to tend to your family, loved ones, and everybody else. Yet it is the same with all living beings and all intelligences, since all living beings and their intelligences are specialized in similar tasks throughout cells, organisms, or in the outside world, in society and in the entire world. While as you notice, intelligences are just as alive as the living bodies, since from physical perspectives they are physical bodies, while from cognitive perspective they are intelligences. It is similar with computers, since from physical perspectives they are the hardware, while from inner subjective digital perspectives they are the software.

It is not too hard to change perspectives, while you must always change perspectives in order to follow your topic of study. Therefore, every time you study thinking, reasoning, feelings, and cognition in general, you can switch to cognitive perspectives if necessary, in order to understand everything better.

Right now in our study of the human origins, we must switch to cognitive perspectives in order to understand how living beings are capable to cope with the environment while forming entire species and classes of life alongside other living beings, and how they are capable to develop themselves through entire successions of species during gestation or during the first stage of life. They certainly know well how to do everything, but we want to know how and why.

The current science states here that all living beings know well how to read blueprints, since the DNA and RNA contain all the necessary information helping all living beings develop themselves from birth to adulthood. Therefore, we must know everything about DNA, RNA, proteins, enzymes, and the entire genetic process of transcription and translation in order to understand growth and development, since this is a developmental physical and cognitive ability.

This is a subconscious developmental ability, yet as you notice, it is a cellular ability, made possible by the first cell, as it knows well how to develop itself right after conception, and how to divide itself in a multitude of cells in order to form the entire organism. From cognitive perspectives, the cellular developmental intelligence knows well how to divide itself while forming the entire organism in the slightest details. While from social perspectives, all cellular components know well how to work together in a harmonious specialized manner in order to make the development of the entire organism possible. Because from social perspectives, cells are societies or civilizations of living beings, while every time cells secede, they form colonies of living beings who grow to form new cells similarly specialized, containing the same intelligences that the parent cell did, in an intact form. Intelligences never die. More precisely, the current intelligences still around had never died, while the rest died, with some intelligences still around since the dawn of life.

While now you can combine the physical perspective following specialized cells and specialized cellular components as these make possible entire specialized intelligences that work together by specialization while helping each other while fulfilling everything within cells and in the entire organism, since all cellular intelligences expand outside cells while forming wider systems of intelligences together by specialization meant to tend to the entire organism, while forming in this manner the primal subconscious intelligences already seen, as the eating primal subconscious intelligence, reproductive primal subconscious intelligence, or the recovery primal subconscious intelligence.

What exactly are genes, if they are not the blueprint of living beings in general? Genes are physical bodies as seen from an objective perspective, while they are intelligences in themselves as seen from an inner, cognitive perspective. These are common intelligences, which are very similar to you, the conscious intelligence, at your own cognitive level.

In order to understand this statement, you have to understand how proteins are formed in large numbers through the processes of transcription and translation, while you have to understand the entire cellular activity as it takes place normally, continuously within all cells, since this is how you understand cellular life, organic life, and therefore living beings, including living human beings.

Proteins are alive, since everything composing life is alive. All proteins are specialized, and they are meant to perform various specialized tasks throughout cells by the zillions, just the way cells themselves are meant to perform various specialized tasks throughout the organism, and just the way human beings are meant to perform various specialized tasks throughout society. Because life is specialized everywhere, throughout all forms of life and at all levels of existence. Once you understand these, you understand life. Because everything is alive, while every living being is composed of a multitude of smaller living beings, performing all inner tasks.

Throughout the world, people build ships and bridges whenever they need them, while using various tools, schedules, schematics, and blueprints in the process, doing the same for all cars, houses, clothes, and chairs. However, all living activity within cells, organisms, and organelles is different, since all activity is alive. Every time the cell must package or transport materials, or every time it must intake ions or build microtubules, it does not simply choose the right blueprint, schedule, and procedure to do everything, but it makes in mass the proper specialized proteins and enzymes capable to do these jobs themselves, proteins that already know how to perform these tasks well. Because this is the only thing that these specialized proteins know and do, they can perform only these specialized tasks, and they perform them perfectly, for as long as they exist.

Because the DNA and RNA do not contain blueprints or information, but they contain entire prototype proteins whose intelligences know well how to do all specialized jobs. However, if they cannot do them as they fall outside their own specialization, they know how to send needs and feelings to all the other specialized protein intelligences to help, forming in this manner very large systems of intelligences that know well how to tend to all specialized tasks within the cell, within the organism, and in the outside world within society and within the entire world. While in this manner, many specialized intelligences cope successfully with the environment in the outside world, since this is their own specialization.

Notice how the existential human arrow points within, and now you have to understand cells, cellular organelles, and especially proteins and amino acids within cells, along with their activity, origins, development, behavior, and entire cognition. Since the more our study anchors on everything, the more accurate it becomes.

Proteins are the little workers that you see from an objective perspective, while they are protein intelligences in themselves as seen from cognitive perspectives. It is similar with the human brain, since the human brain is actually the human mind as seen from a cognitive perspective. While you are the conscious intelligence from a cognitive perspective, not the entire cognitive system.

To make matters more complex, at their very small scale,

proteins lose shape, rigidity and even material consistency, because everything is field at their very small scale. You cannot even distinguish objective details anymore at their small scale, because visible light itself has wavelengths larger than the small size of proteins and amino acids. Therefore, it might even be easier to perceive and understand proteins, enzymes, amino acids, and all molecules composing these from cognitive perspectives as intelligences, as all inner intelligences that these form, carry, and maintain throughout life.

As a reference, every time you study the human cognition including reasoning and perception, you study the human mind along with all human intelligences inhabiting it, more than the human brain holding the human mind, since the human brain is the real, objective perspective of the human mind, while the human mind is the cognitive perspective of the human brain. The human brain is capable to hold, form, and maintain the human mind within itself throughout life, since this is how you are mind, body, and soul, as one.

There is more to consider, since the entire human physical body including the human brain holds the human mind and all its intelligences, conscious and subconscious. While you also have to find the human cognitive origins and development, the origins of all human intelligences.

Since the existential human arrow points within, we should continue our study of the inner cell. How exactly do proteins know how to perform their specialized tasks, once they are produced in mass by the cell to tend to their specific inner cellular tasks while fulfilling specific cellular needs? Notice how all inner cellular needs fulfilled by your proteins correlate and intertwine with your own needs that you fulfill in the outside world as a living human being. All intelligences are systems of intelligences, having at their base zillions of protein intelligences and cellular intelligences, all interconnecting harmoniously by specialization while fulfilling all their needs and meanings, while in this manner all tending to all possible tasks within the organism and in the outside world, undergoing very important meanings assuring the human survival, the human subsistence, and the human development, taking place continuously at all levels: subcellular, cellular, human, social, national, spiritual, and higher. While as you notice, everything is made possible by the multitude of proteins and protein intelligences found at the roots of all specialized systems of intelligences spanning the organism and extending in society and in the entire world, while forming Life altogether alongside all living beings doing the same.

I tend to refer to all life everywhere and continuously as Life, since she is alive, containing you as a living human being just the way you contain the multitude of your proteins throughout cells.

The inner intelligences of all proteins are specialized, and they know well how to perform their own cellular tasks, just the way your own eating subconscious intelligence knows what food to request from you to eat today, when, and in what amounts, sending you all your relevant, distinct needs to eat, with the highest accuracy needed to feed all your cells and cellular components exactly as these need, with the precision of nanograms, while this is what you always eat.

Therefore, once the cell has at its disposal anything to be done within the cell or at its proximity within the organ or organism, it simply produces a specific number of proteins to do the job, and the proteins already know what to do and how to do very accurately. They know it because they already contain the necessary specialized protein intelligence who do everything, just the way your brain contains the mind to do everything necessary throughout life, making you a successful individual living being when you do everything right.

You must have a physical body along with the intelligences that it contains, together forming life, or together being alive. This is how the human brain forms, holds, and maintains the human mind. More precisely, the entire human physical body holds forms, and maintains the human cognition, cognitive system, or mind, since it is the same.

Yet you are more than mind and body, because you are an entire lifeline of existence and not only your physical body and the intelligences that it holds while forming your mind, since you are your mind, body, and soul, at least, forming your own lifeline of existence.

All proteins are created directly in the form and image of each gene found within your genetic material, and this is very vast. We also notice that life is lived at cellular and subcellular levels, as these gather within organisms and societies to live life as larger and larger forms of life, probably in order to think better, but always in order to be more fulfilling, more developed, more capable, and more successful. This is why you have a cellular nucleus in each cell and not some centralized genetic organ to offer the blueprint for the entire organism.

Therefore, your genetic material is not the blueprint for life and for the human being, but it is a place where all prototype proteins live, called genes by the current science. By producing proteins in mass, you form all proteins and enzymes in all cells, proteins that perform all specialized tasks within cells. Many of these specialized cellular tasks performed by all proteins tend to the casual physiological cellular activity, while some cellular tasks regard directly the cellular, organ, and organism development, reproduction, thinking, interconnectivity, and social acceptance, since all needs and meanings that you ever receive and fulfill as a living human being come from within cells, and you fulfill everything just for them. Yet everything that you do in life you do to fulfill your needs, and you do so for your subcellular components, for all your cells, for your entire organism, for your family, for all your loved ones, and for the entire society. This is why you breathe, sleep, learn, eat, develop, and reproduce, for your cells, to feed them, to develop them, and then to make sure that they survive after you die, through your children and through their children, throughout your genetic line and as long as your genetic line exists.

Why having life developing in this extraordinary manner from one form of life to another, from amino acids, to proteins, to cells, to organisms, to societies, to civilizations, to realities, up to Life herself? It seems that the entire model of life provided by science is too simplistic and therefore it cannot help us much throughout our research. We can study gestation right now, to interpret it as a genuine evolution from the single cellular life of the distant past to the current developed species and then to humankind. While it seems more probable that life does not actually have to die with each living being and then to evolve all over again with the birth of each new living being, but there is a continuation of life throughout all living beings, and this specific continuous living presence found in all living beings has to develop shortly during gestation, up to where it had been only one generation previously.

As you notice, life is not exactly about organisms and entire species, but life takes place at the cellular and subcellular levels, and further below. As we notice, life originates at the cellular level and below, only to develop throughout the first stages of life of each organism as a union of cells, with the cells themselves or with only their cellular components animating and giving life to the entire organism.

Yet it is only a matter of perspective to state that the entire organism is alive, because your senses of perception and therefore your perspectives are set on the entire organism and at the level of your entire organism. You are the entire organism as seen through your own senses of perception, yet when you study the entire organism in component details, among the entire organism and among the entire cognitive system, you can identify yourself only with what you are capable to influence directly. This is how, from the entirety of your organism and cognitive system, you are the conscious intelligence, and nothing else. More precisely, you are one of the multitude of specialized intelligences of your cognitive system, as your eating, recovery, or reproductive subconscious intelligences. While you the conscious intelligence are specialized and responsible with the coordination of the entire organism as it interacts with the outside world while it fulfills its needs. Or while you fulfill your needs and meanings as a conscious intelligence, since your own needs and meanings are actually the conscious needs and meanings of the entire organism in the outside world.

You notice that you can relate life with the organism or not, or with your cells or not, or with the entire humankind or not, with the entire ecosystem of Earth or not, or even with your cognitive system, with your intelligences, or with your own conscious intelligence, or not. Because life is at all levels and classes, and it is in each one of your components, as cells, organs, cellular components, or intelligences of your cells and entire cognitive system. Even more, we can define life as being at all these levels separately and simultaneously. While for many humans, life goes far above this world and far above the human condition here in this world, because you must always have a body and an intelligence animating it in order to have life, and therefore you can have these at all levels, classes, and forms: molecular, cellular, organic, social, and higher.

The current science defines life through its main empiric characteristics, as moving around, surviving, and interacting with the environment. Yet the current science defines only organisms as being alive, probably because human beings seem to be only organisms, therefore humans are alive, and therefore only organisms are alive, or this is what the current science states. While studying human origins, along with the origins of life, we notice living beings that are not only organisms, and even living beings that are not even cells. What is life now? Is it organisms, cells, or cellular components?

We notice that life takes place at the cellular level, and then it only unites cells by the trillions to form entire organisms, for survival, subsistence, development, and efficiency purposes, just the way organisms live life together throughout schools, packs, herds, cultures, societies, and civilizations. I refer to these as forms of life.

When we study the origins of the human mind and of the human cognition in general, we notice life taking place at the level of the cognitive system, with you being the conscious intelligence as you inhabit this physical body. All intelligences span cognitive systems that inhabit physical bodies, or you can refer to this as minds inhabiting brains, with all intelligences uniting to form entire minds, just the way cells unite to form organisms, living life in form of organisms. While everything happens at the cellular level, with cells having their own cellular intelligence, intelligence capable to unite with the rest of the cellular intelligences forming entire cognitive systems as the human mind. Life is the union of the body and mind, the union of all cells, the union of all intelligences, all in one, all alive, and more importantly, all alive at all levels: cellular, organism, species, social, and higher.

For example, you are the entire body, while you are the mind. Yet you are not capable to access your entire mind, but only your consciousness. Because your subconscious main intelligence with all its subconscious inner intelligences is out of your conscious limits, and this is why you cannot digest your food on your own consciously, you cannot control consciously the filtration taking place continuously in the kidneys, you cannot control consciously the movement of your stomach, you cannot do too much as a conscious intelligence within your organism but only in the outside world, since this is your specialization. Other intelligences do other tasks throughout the body and mind, you call them subconscious intelligences since they are not you and you cannot access them directly, and this is who they are and what they do.

Who are your other intelligences? Your mind or cognitive system divides into three main intelligences: conscious, subconscious, and highconscious. All main intelligences are formed of primal intelligences, and you as a conscious intelligence are one of them. While all intelligences have a body together, the physical body. You have mind, body, and soul. You have your lifeline of existence capable to define itself, and this is the definition of life. Yet among all your primal intelligences, as your eating, recovery, social, security, developmental, or reproductive primal intelligences, you the conscious intelligence are the only primal intelligence found on your own lifeline of existence, with all other primal intelligences of your cognitive system having their own lifelines of existence.

You cannot refer to the physical body as being alive without the mind and the multitude of intelligences controlling and animating everything, while you cannot refer to intelligences as being alive without the body itself. It is not a coincidence that the human body, the human cells, and the human mind originate at the same moment and place, because they are the same living being, as one. Because from an objective perspective placed here in this world, you are the physical body. While from the inner, subjective, cognitive perspective of your mind, you are the interconnectivity of all your intelligences, you are the cognitive system or the human mind. While you are alive everywhere you are. You are alive objectively and you exist objectively in the outside world, and you are alive subjectively and you exist subjectively in the inner world of your mind. You exist there as your inner self, which is the exact replica of yourself from the outside world.

Notice how living beings are not only physical bodies as depicted by the current science, but living beings are mind and body as one, living life in this manner in two distinct realities, one outside in the real world as physical bodies, and one inside the mind as conscious intelligences. This is why we must always use different perspectives for different worlds and different selves, because all life is lived simultaneously in many realities. While in order to be in other realities, you must have a self of yours already there. You have the physical body here in the real world, you have the conscious intelligence or inner self in your mind, and you have your soul in the higher world.

Notice how all worlds and realities where you live your life are not random, but they are one within another within another, forming in this manner a line of existence. The higher world holds this world within, while this world holds all inner mind worlds within, and in this manner, all inner mind worlds exist within this world, which exists within the higher reality, which exists within even higher realities, up to Life herself and the wider world.

As stated, in order to live life in any world or reality, you must have as self of yours already there. Yet since all your worlds are lined up one within another within another, your selves form an entire line of existence, since all your selves live life one through another through another. In your case, your soul lives life as the conscious intelligence or intelligent inner self, which lives life as the physical body in the outside world. Yet your soul can live life only in his higher world if he wants, while you the conscious intelligence can live your life only within your intelligent mind spanning the cortex if you want, while reasoning, planning, or daydreaming, never interacting with the outside world. Yet if you want, you can open your eyes and walk around in order to interact with the outside world, yet you do so as the physical body, since your physical body is the self that you have in the outside world. While you always feel, reason, and daydream as a conscious intelligence.

You do not reason as the physical body, because you can reason only in your mind world. Similarly, you cannot interact directly with the outside world as a conscious intelligence, but you must interact with the outside world only through your physical body, only by embodying your physical body. While in this manner, the conscious intelligence is the avatar of your physical body. Yet it is more likely that your soul embodies you the conscious intelligence that embodies the physical body, forming in this manner your own lifeline of existence that comprises all your selves: your soul on top, your conscious intelligence in the middle, and the physical body on the bottom. While in this manner, your soul is the avatar of your conscious intelligence, and through it, your soul is the avatar of your physical body, and this had always been the case here in this world ever since this world had been created. As a reference, the word 'avatar' is among the oldest words in use in this world, helping you now identify the origin of this entire world.

Your higher self precedes all your selves from this world, since it lives in our upper reality. It is similar for its own souls, since these are your higher selves just as well, preceding your own soul on your lifeline of existence.

Yet you have other selves here in your physical body,

within specific cells and areas of your brain and body. While all your selves live life one through another if they choose so, to be your conscious intelligence, then your inner self from the cortex, and then the physical body if they choose so. While you have your social self in the intelligent human society, currently missing, and your consensual self in the Consensual Matrix, which you are most of the time or continuously, even unknowingly. While as stated, all selves of your lifeline of existence have different lives, existences, meanings, fulfillments, and therefore origins, while your lifeline as a whole, if it is divine, it has the same origin as Life, Intelligence, the Divine, or the One.

Notice the multitude of souls, spheres of existence, realities, forms of life, classes of life, and natural and consensual interconnectivities that we have to follow throughout our quest of finding the human origins, because human beings are very complex, in a very complex world, throughout a very complex life, and we must consider everything.

By studying the development of any living being throughout gestation, we are able to study the development that organic life undertook throughout time, and this is part of our study of the human origins and of the origins of the organic form of life. We notice the development of the physical body from an objective perspective, and the development of the mind, cognitive system, and all its intelligences from the inner, cognitive perspective. You notice the specific part of life that does not die with each generation but it is transferred wholly and alive from one generation to another. It is in you now, but you have transmitted it to your children, going from there to their children. These are your primal intelligences, conscious and subconscious, alive now in you ever since the dawn of life, since they had never died.

Therefore, your developmental subconscious intelligence had been there at the dawn of life, and then it transitioned successfully from one species to another alongside similar living beings for a very long time while following all major changes in the environment and while defining your own genetic line, until it reached the human species from generation to generation to be you now as you read this book. This is why, once you are conceived, your own subconscious developmental intelligence must transition fast during gestation from one species to another in order to form you in your human form, because this is the only thing that your subconscious developmental intelligence knows what to do, since this is how it developed gradually itself ever since the dawn of life, while passing through all species and classes of life along your own genetic line, and this is what it does, this is its own specialization within your subconscious mind. While as you notice, it is significantly easier to understand the origins and development of life and of humanity from cognitive perspectives. While the current science will never understand anything about life and humanity only from physical perspectives and only by using genes and taxonomy while always ignoring intelligences, since it is impossible.

Your subconscious developmental intelligence is formed of a zillion protein intelligences at its roots, protein intelligences that know very well how to shape you in all physical and cognitive details, because all these protein intelligences were right there throughout the development of the entire organic life coping successfully with all conditions of the environment, while giving you the multitude of your abilities and characteristics making you be, look, and behave exactly as you do. Yet this is the case with all living beings, since all their developmental intelligences behave in this manner.

However, you must always remain aware that the same subconscious developmental intelligence developing the entire human species, developed you as you are now since birth, while still developing you in every meaningful manner through all developmental needs that it sends you, that you receive as a conscious intelligence, and that you must fulfill in the outside world, in order to help you and your family, loved ones, and everybody else cope with all significant details of the environment, in order for you and the entire human species to survive, subsist, prosper, and continue developing, alongside all living human beings, and alongside all living beings of all species from this world. While this is how all living beings know exactly how to cope with all details of their environment, because they always receive needs from very pertinent subconscious intelligences to do so, not only from developmental intelligences. While in this manner, notice how all living beings work very hard to fulfill all their needs, including developmental needs, doing so mostly as conscious intelligences, never fulfilling them randomly or accidentally as the current science implies.

While even this specific need for intelligent learning that you fulfill right now while reading this book is part of your developmental needs, while as you notice, your intelligent developmental fulfillment is neither random nor accidental, because you work very hard to find these books and read them, while always reasoning intelligently alongside them in order to identify, understand, and elaborate all important knowledge from this book. While this knowledge or only part of this knowledge helps you in the future throughout the fulfillment of all your needs, including developmental needs, helping you in this manner to cope with all changes from your physical and social environment individually or as an entire species.

This entire complex development might seem implicit or only optional to you, since you always have a choice throughout the fulfillment of all your needs, including your developmental needs. However, since your subconscious developmental intelligence was born at the dawn of life, it is now not only in you and in your entire genetic line intact, but it is in all human beings in an identical manner, together sending everybody similar developmental needs, which are mostly harmonious, determining everybody to fulfill everything together harmoniously. While this is the case with all living beings, since all developmental intelligences are similarly in all species, but only in a relatively identical manner, yet still sending all living beings harmonious developmental needs and meanings, while determining a harmonious behavior in this entire world.

This is how your own human origin stands at the moment of your conception, while the human origin of all humankind stands at a distant moment in the past yet still correspondent, because the same developmental intelligences and all your primal intelligences were there back then, and are here with you now. The origin of humanity coincides with the origin of your own human intelligences, while this is very important to consider, since it makes your existential arrow point inside you, towards your proteins, amino acids, and towards all molecules and atoms composing them. Then continuing further inwards, pointing towards your entire mind, then towards main intelligences, primal intelligences, inner intelligences, cellular intelligences, and protein intelligences, all the way down to the raw electromagnetic field, since this is only the electromagnetic type of life. Yet your cognitive and spiritual world is not bounded by the electromagnetic field of this world, but it continues from there to your higher and higher realities, as far as your lifeline of existence lasts, because souls always have souls that have souls, and you never know how far you reach.

Let us continue this specific higher existential arrow in the following chapter, as it intertwines with your living and consensual existence in order to create this comprehensive human experience.

2 COMPREHENSIVE HUMAN EXISTENCE

I used above terms as existence, life, and reality in the same context. Do you actually exist, or you are only an illusion, as in an extraordinary videogame, without even knowing it? Because if this is the case, if this world is a created reality, mostly an artificially created reality, or even a digitally created reality, then our current topic is very easy to define, since it had been created in this manner including humans, human intelligences, and human origins, to be exactly as it is today, or to develop in this exact manner or in any manner desired.

Does this world exist? Does Life exist? Do minds and intelligences exist, when you cannot even see them?

I refer to Life as the entirety of all living beings everywhere, in all forms of life, continuously and throughout all realities, and therefore throughout the entire wider world. I refer to Existence as the existence of Life herself. Therefore, Existence defines Life directly, while it also defines all living beings composing Life including all human beings, and including you. You are alive, you are part of Life, and therefore you are defined by Existence through Life. You exist through Life, while your own existence defines you, directly, here on Earth.

Existence in general has three natures: objective, subjective, and highjective, all defining Life comprehensively, in her entire living harmony. While all agreements made among any living being exist separately from the Existence of Life, defining only themselves, throughout their own ideologies and jurisdictions, and these are never part of Life. These are only agreements that they exist and are true and valid, and they only validate themselves in an abstract, unnatural, unreal, nonliving, consensual manner, with Existence ignoring them altogether.

As a reference, you can make any agreement within your entourage, that summer nights are the best, or that the sky is actually unreal. While through similar agreements, you can validate to be true everything that you happen to decide. Therefore, if the entirety of employed scientists of Earth decide that the big bang actually took place and formed the entire universe and they validate it themselves to be true, then this is an example of consensual agreement, belief, or even law if they decide to state it in this manner, yet it remains valid only throughout the jurisdiction or ideology of the current science. You also have to agree with them in order to keep your job or to pass your exams, yet the entire Big Bang is not necessarily accurately true and you cannot even tell anymore, mostly if you have never had the chance to witness it yourself.

But did the big bang actually exist? Because if the big bang was an actual event of this world, then it certainly took place, while it defines the origins of this world, always validated through the Existence of Life. If this is the case, then it is an accurate truth, since it was an actual event. Yet if the big bang was not an actual event of this world, then it is only a consensual truth, as it is currently validated legally, consensually, and even unanimously by the current science. However, it seems that the current science prepares to discard this consensual agreement, the big bang itself, in order to adopt something else, a steady state model of the universe. While the current science can always do so, through yet another agreement, stating that the previous agreement is false, and therefore it is no longer consensually valid. While all these agreements have nothing to do with the accurate truth of Life and of the real world.

Everything here in this world exists in an objective manner, as all objects, air, water, human bodies, dust, and electromagnetic radiation. Similarly, everything within inner mind realities exists subjectively and cognitively, while everything throughout higher realities exists highjectively. Because existence is Boolean while distinguishing the existent from the nonexistent in a trivial manner, while existence is also relative, through the law of relativity found at the base of this world, defining everything in a relative manner subjectively, objectively, and highjectively.

In contrast to the accurate truth, which is always accurate, agreements are made among any living being, with most of them in the mainstream. While if you are also mainstream, through your party, lodge, ideology, regime, jurisdiction, or employment, then you must agree with the entire mainstream, otherwise you get in trouble. Because otherwise, if you do not agree with them, you go against the law, and you suffer in consequence. Everybody knows this, yet as you notice, you always get in trouble by agreement, suffering in consequence by agreement just as well. While they have to instate consensual, agreed courts and jurisdictions in order to display their pertinence to judge and to convict you, if you only happen to go against their laws and agreements. Because authorities define their own courts, jurisdictions, laws, agreements, beliefs, norms, and entire ideologies as valid and therefore as good and pertinent, and if you go against these, you get in trouble. While as you already know, many times, authorities are only the most powerful organized criminal gangs among the rest, deciding now everything that is good in the world by agreement, by their own agreement, on their own behalf.

However, in order for everything to work fine, all these consensual authorities must agree with each other throughout nations and throughout the world, forming in this manner a genuine main, valid, mainstream, consensual matrix, containing all good, legal, pertinent and lawful jurisdictions, parties, and ideologies of the world. Yet with all mainstream consensual matrices of all worlds and realities remaining in agreement and validating themselves through their own consensual agreements, laws, and ideologies, this extraordinary consensual matrix spanning most of the wider world is unique, and it is called Consensual Matrix. While you either agree with it, or you are against it altogether, and you getting in trouble.

It is important to understand the Consensual Matrix, because the Consensual Matrix engulfs Earth entirely, along with its current science, medicine, finance, justice, education, and entertainment. Consequently, everything that you do, learn, and desire in this world is consensual, by mainstream agreement, and never part of Life, truth, and reality. Because if the Consensual Matrix decides that the theory of evolution, the big bang theory, and the theory of relativity must be true this century but not the next, then this is what you must learn, understand, and teach as valid knowledge.

Many people and groups of people can agree on everything they desire throughout the world and everywhere else, defining anything to be true in this manner, but only consensually true, or legally true, or ideologically true, all being consensual. We already know many of these theories defining anything including the human origins, but we want more, we want the actual accurate truth.

We are never against anyone by seeking the accurate truth, since we are never against the people instating all consensual truth in the world. We respect everybody, as long as they remain within the mainstream, and many times even in the alternative, since the Consensual Matrix accepts many times even the alternative, being part of its consensual self.

Because as you notice, the Consensual Matrix engulfs justice as it is instated here on Earth, while justice is still helpful throughout society and throughout the world, keeping you safe. This is the case most of the time, depending when and where you live.

It is stated in the ideology of the current science that it seeks to learn and explain all truth, without mentioning that this truth can be either accurate or only consensual, validated by agreement, as it is the case with all consensual knowledge. While we always seek accurate truth, not consensual, since we already know all theories and beliefs related to the human origins even from school and from the entire society. We want the accurate truth. We want to know exactly what took place exactly then, at the beginning of the humankind, at the beginning of life on Earth, at the beginning of the organic form of life, at the beginning of Life herself, at the beginning of this world, and even at the beginning of the wider world.

We already know the multitude of theories, beliefs, and even stereotypes defining the human origins. We seek to know which one is true, accurately true, and if none is accurately true, we seek to find this truth on our own, if it is ever possible. The problem is that everything happened very long time ago, and we were not there to witness it ourselves firsthand. Because everything that you witness firsthand objectively is objectively true and objectively accurate, and this is the strongest truth ever. Similarly, right now, you remain incapable to go past the upper orbit of Earth in order to witness the actual shape of Earth. While all records coming from the current science are mostly consensual, and even those claimed to be taken objectively, empirically, firsthand, are not convincing.

This does not mean that the Earth is flat, as many ideologies and jurisdictions already state, but the problem is that the current science is consensual, and now by claiming that the Earth is spherical, it does so consensually just as well. While you cannot learn the accurate truth about the shape of Earth from here at the surface of Earth, because you must go in the orbit of Earth to see it yourself. You cannot even calculate it or assume it in any manner, but you must see it yourself, you must go in the orbit of Earth or further out in space to see it yourself. Which is currently possible only as an astronaut and only if all space missions are not fake. You can also study all records from all space missions, and if you find them conclusive, then you can state that the Earth is spherical. However, as you notice, even this is not accurately true, but only valid to be true, as you validate everything yourself. While the current science had already validated all space missions as true, yet even this is only valid truth, not accurately true.

However, in all space records you never see the Earth, but only some blurry images of Earth. There are only seven of these left on the Internet, while as you study them closely, you find them fake. There used to be more, yet people pointed out all possible inconsistencies, and the authorities removed them, leaving only seven behind, which are still inconsistent. Yet this does not mean that the Earth is flat, but only that space missions never take place.

There are still intelligent means to find out the shape of Earth even from here from the surface of Earth. All land on Earth is owned, while all land measurements are calculated very precisely. When you combine all land, you obtain exactly the sphere of Earth, with the exact shape and size offered by the current science. However, if this world is as large as the lower orbit of Earth, then the Earth itself does not have to be spherical but in any manner, even flat, since its shape does not matter in a world that is slightly larger than Earth.

Yet once you believe otherwise, that Earth is not spherical, that Big Bang never took place, that life had never originated here on Earth, or that humans do not come from monkeys, you might still not be right, since you are part of the alternative science, while even the alternative science might not be right. Yet as you look around, you can see the alternative as part of the mainstream, since it is controlled along with the mainstream through the Consensual Matrix, with everything else considered by the same overall agreement as illegal, erroneous, immoral, and even grotesque or impossible. The alternative is still in the Consensual Matrix, since the Consensual Matrix is capable to control the alternative science just as well, as you can easily tell by searching the Internet.

What is this Consensual Matrix exactly? Is it the devil? No, but it can become anything that it wants to be, including the deity itself, through agreement that it is the deity itself. Since this is mostly the case, but not the devil. The Consensual Matrix is a genuine reality in itself, only that it is consensual, abstract, and therefore not part of Life, standing outside Life and the wider world altogether, by agreement. The current system of justice from Earth is integral part of the Consensual Matrix, along with all legal cults, ideologies, business structures, lodges, jurisdictions, and much more. This might be everything that you know of the Consensual Matrix if you are from the Brotherhood or the Masses, while you might have to interact directly with the rest of the Consensual Matrix if you are from the Elite.

And it is not utopic or incredible, it is never hidden either, while it always tries to be fair and helpful. Since

The Consensual Matrix had been instated in the wider world to help intelligent life interconnect in a just, fair, and efficient manner. Within it, you undergo your existence as a consensual self or consensual corporation, brand, or trademark, but you do so only within specific, consensual, agreed circumstances, every time you identify yourself with your name written in uppercase letters from the multitude of your documents, as you do when you conduct business, when you use money, at work or at school, at the clinic, at the lodge, and within courts, jurisdictions, public places, districts, cities, ideologies, and hierarchies.

Therefore, now, if you seek to know, learn, and understand the human origins through laws, beliefs, and ideologies, you remain in the Consensual Matrix, since you might not be able to escape the current science, ideologies, education, media, entertainment, and academia. Yet once you are in the Consensual Matrix, you already know your answers, since this is mainstream knowledge. If this is not satisfactory for you, since you can always see through all lies and now you just want to know the truth, then you can study the alternative science instead, since this is always willing to provide you with the alternative truth: aliens, angels, demons, miracles, psychics, souls, ancient aliens, vampires, and holographic universes. However, if even these seem inadequate to you, as they do not even remain conclusive among themselves, then you have to reason independently, seeking the entire accurate truth on your own.

Additionally, you have to distinguish between the living and the consensual, and therefore between Life and the Consensual Matrix. Because only in this manner, your reasoning remains accurate, based on the natural laws of this universe and of the wider world.

Yet can you actually distinguish between Life and the Consensual Matrix? Can you do so even here on Earth? Because as stated, most of the people live life in the Consensual Matrix continuously, willingly or unwillingly, knowingly or unknowingly. Is this your case?

Who or what exactly are you? Everything depends on the specific existence defining you. Because through the Existence of Life, you are a genuine living human being, and you fulfill your normal, natural needs, for your life and the world, and through you, for Life and for the wider world, including your animal needs, physiological needs, social harmony needs, developmental needs, and family needs. While through the agreed, consensual existence defining the Consensual Matrix, you do everything consensually, you fulfill all your consensual needs, and you do so through your name written in uppercase letters, which is your brand, trademark, or corporation. While you exist consensually in this manner within jurisdictions, districts, courts, and ideologies, but never in Life and in the real world. While you have to respect, believe in, admit, acknowledge, and therefore agree directly or implicitly with these courts, jurisdictions, or ideologies, otherwise you cannot be there and they do not want you there as a living human being while you seek truth and knowledge outside their own agreement.

While for them, it is only a charade, just as agreeing that the sky is yellow, since it is nothing harmful, nothing real, but only an act, only a code, only a show, only a play. Until they take away your car and your house, along with the custody of your loved ones, since that is real but it is still assumed consensual, through agreement, only a play.

Existence is always only Boolean and relative while defining the subjective, the objective, and the highjective. These are relative to each other, since existence itself is relative. These stand one above another, making the objective itself to become subjective as seen from above, from a highjective perspective. In this manner, your daydreams and all your intelligences are subjective in nature, as seen from an outer, real, objective world perspective. While you become subjective just as well, along with this entire world as seen from the highjective perspective of our higher world, by all souls and higher beings up there. Which means that you are a higher being as seen from the perspective of your own intelligences. While you as an organism or physical body are subjective in nature as seen from the perspective of all higher beings including your soul or higher self.

Furthermore, if you and the entire Earth are part of an extraordinary videogame or mind virtual reality, you still exist objectively from your own perspective, because the specific reality that you happen to inhabit always offers you an objective perspective, an objective existence, and therefore a normal, objective life. This is why dreams seem so real, despite of all ambiguity taking place there, because dreams are real and objective but only for you and only as long as you are there.

While the Consensual Matrix stands apart from all these, in its own, distinct, valid, pertinent, consensual reality, defined by its own consensual existence through agreement, while still interfering directly with all realities systematically. Because the souls bring the Consensual Matrix here with them every time they come here on Earth, in our world. Yet do not blame them too much, since you take the Consensual Matrix everywhere you go, at home in your family and online throughout all websites, or in your books and in all movies, and throughout all courts, jurisdictions, and corporations that you can ever access or instate, since everything has to be part of the Consensual Matrix as it always claims, demands, and enforces very strictly.

But can it actually be so perfect and necessary? Can Life and Existence itself accept, create, and even demand the

Consensual Matrix? No. Life has nothing to do with the Consensual Matrix, since the Consensual Matrix even gets in her way, even by domesticating her, and by hijacking her intelligent living beings. Because Existence itself can define only the subjective, the objective, and the highjective, while the Consensual Matrix is only abstract, but not subjective as it can state even by law. The Consensual Matrix is instated and maintained only consensually, by the agreement of all intelligent living beings choosing to do so, knowingly and unknowingly, willingly and unwillingly. Which works well mostly in court. Yet since Life always respects the decisions and agreements of all her intelligent living beings, now it works even better. It works everywhere, since this is how you end up in the world with the existent, the nonexistent, and the consensual.

People always knew about the Consensual Matrix throughout the past ages of Earth, only that the Consensual Matrix can always write and rewrite everything as it wants, including the human history.

The old, genuine, egalitarian Brotherhood of the past always stated it clearly: to be or not to be, noting else. Which means the existent or the nonexistent, but nothing else. Yet this was the old, living, harmonious, egalitarian Brotherhood of the past, a priceless remnant of the old, living, egalitarian, harmonious ages of Earth. However, the old, intelligent human Brotherhood had been hijacked by the invisible kingdom only centuries ago, more or less, depending on where or when you live, and this is how now you have the current hierarchic Brotherhood that you know well.

While in this manner, you have the existent, the nonexistent, and the consensual altogether. While through the consensual, the Consensual Matrix has access to Earth. While through the Consensual Matrix, any higher being up there has direct access to Earth, doing everything that you see today and always throughout the current age of Earth. Yet it is the same in the entire wider world, so it must be normal. It is surprising how the old egalitarian harmonious Brotherhood had made it so far in the current age of Earth, with the Consensual Matrix already instated on Earth for thousands of years, while already instated in the higher world for even longer. While everything relates to existence itself.

Your existence is objective in nature everywhere you are and for as long as you are there, but only from your own perspective. Because existence itself is relative, and in this case, it is always relative to your own perspective. Because from other perspectives, others can perceive and understand you differently, depending on the realities where they are. You can be only subjective in nature from the perspective of the higher beings interacting through you throughout the extraordinary virtual Earth videogame from our example. You still exist, Existence still defines you objectively here, yet it defines you only as a subjective living being from their higher perspective, similar to any intelligence and videogame character.

There is a difference between intelligences and videogame characters, since intelligences are natural, always part of Life herself, while all videogame characters are subjectively artificial, created artificially by you or by others. Book characters are subjectively fictional, while your own corporation is subjectively consensual, defined by law and with stamps, signatures, and registration numbers. Therefore, throughout life, you identify yourself only through your own corporation, and this is how you live your life, consensually. This is a fact, because you have all your documents and cards of identity to prove that you are a corporation and not alive, and therefore that you have the rights and status of any consensual corporation, which are very low. This is why anyone can own you and use you in any manner, just as you own and use objects and tools. While it cannot even be called exploitation, but simply ownership, just the way you do not exactly exploit your car or computer, but you simply use these, as property.

In an entire consensual society, your origin is defined exactly by your documents, since all ownership certificates of all corporations have their date of birth inscribed on them, and that is their origin. Or that is your origin, if you identify yourself consensually, through your own corporation. You can also notice a name written in uppercase letters there, while that is the brand, the corporation itself, not you the living human being. It is not your name as you might assume, and it is not even the name of your corporation, but it is the actual brand, trademark, or corporation itself. While many corporations can be anything else, as any name or symbol, or as the flag of a nation.

Whatever the case, corporations are considered dead, they are not considered objectively real but only consensually real, which means that corporations exist through consensus alone. Therefore, corporations are not part of Life and they are not part of this world or of any reality that is part of the wider world, since the Existence defining Life defines the wider world in all its realities, including this world, including all living beings that are part of Life and therefore including you, along with all your intelligences, yet existence and Existence never define anything consensual. Because Life is the life of the wider world, and Existence is the existence of Life and of the wider world. While all consensual corporations are by agreement, they are based on legal, consensual truth, and they do not exist objectively in this world and in the wider world. To be or not to be, but nothing in between.

Corporations still have to exist somewhere, otherwise, even the consensual existence cannot define them. Yet since humans consent to their legal, consensual existence, these corporations including yours exist in consensual worlds called jurisdictions, or districts, cities, or states. Even countries and nations are only consensual in nature and therefore owned and legally exploited, since the statute of many jurisdictions define countries, cities, nations, and even persons and drivers to be consensual, only corporations, as they are incorporated first.

This must be a conspiracy theory or only a cheap trick, since this is what you think if you are from the Masses, because the current hierarchic Brotherhood already know it or know part of it, having a share of the profit.

This is the case on Earth for a very long time, with the

word 'corporation' coming from the old Latin 'corpore,' meaning body, dead body, voided of life, or absent of life. Since this is the meaning of religion, to manage the bodies or corporations of all living beings here on Earth, with the Divine managing the souls themselves, and with all living beings living their lives freely here in this world. While not too many people know these.

Why are people exploited, discriminated, and even eradicated in this world? If they are free, then why are people still obeying the laws, and why are they still judged, charged, condemned, and punished, if they are free, and if none of these should apply to them? If you are from the Brotherhood, you already know the answer, since you are made continuously to assume and represent your consensual corporation. You are not you the living human being, but you are only a fiat corporation, through your own agreement as a living human being. Since all living human beings should live life in any manner they please. While corporations are owned and therefore controlled tightly today and always, probably ever since the human origins. Or this is the case if the Consensual Matrix has planted humanity here on Earth and in the world. This is the consensual purpose of this world, servitude, discrimination, exploitation, and incrimination.

With the Consensual Matrix profiting from everything, while the Consensual Matrix is consensual in nature, it is only an overall consensually legal reality or jurisdiction spanning most of the wider world, and therefore standing on top of Life, now represented by a larger part of intelligent life of the wider world. With you, the Brotherhood, the human society, the human civilization, and with a larger part of the world actively involved. The Consensual Matrix decides how much you know and understand about yourself and about your life and your world, including the human origins and the human development.

This is what science, education, and the media teach you, and nothing more, while hiding the rest of the knowledge on purpose, and while getting in your way every time you seek to know it. Because once you know it, you might choose not to be part of it, you escape the Consensual Matrix, and you live your life freely, as a genuine living human being, in your own real world. While you can even achieve to live a successful, fulfilling life at the intelligent human level in this manner, going even against the Consensual Matrix. If others side with you, you manage together to create the intelligent human environment, and this is not part of the Consensual Matrix.

The human beings managed to remove the Consensual Matrix from Earth. Or they actually came from beyond and moved here on Earth in order to escape the Consensual Matrix. They lived a highly successful life here, developing to their intelligent human level, while recovering all their higher cognitive abilities. Yet only thousands of years ago, the Consensual Matrix infiltrated and destroyed them. Even today, the current hierarchic Brotherhood still works hard to remove the old, free genetic traces on Earth, since this is what the ongoing terminal illnesses, pandemics, genetic alteration, and genocide are. With the current consensual Brothers working overtime to implement them, eradicating the old genuine Brothers along with the Masses and everyone who is not already consensual.

Why are you a subjectively consensual corporation here on Earth, despite the fact that Earth is objectively real and so you are real, natural, alive, and free, as a normal, natural living human being? Even more, all accurate natural laws of the Universe define you to exist objectively here in this world, because they define you both as a material being and as a living being. You are also intelligent, this is also a fact, yet you are a consensual corporation only through choice and agreement, at your own request, and you are in this manner ever since you have applied for your certificates, licenses, and the social security number or social insurance number that you use now in order to find employment, earn money, and use it throughout society in order to fulfill your needs.

As stated, existence has three relative natures in the wider world: subjective, objective, and highjective. All realities exist only in any of these existential forms, since existence cannot define them otherwise, and they just cannot exist, or at least not for you, not for Life, and not for the wider world. Living beings and intelligences inhabit these realities not necessarily individually, but mostly throughout entire environments, societies, civilizations, worlds, cognitive systems, and ecosystems, depending on circumstances, and these are many to consider. To make matters even more complex, living beings and mostly their intelligences are capable to form or create entire inner realities themselves throughout life, or only instantly throughout casual cognition. This is how, living beings and intelligences are capable to inhabit a multitude of realities of all existential natures even simultaneously, and this is the case with all humans.

Since the Consensual Matrix defines humans and humanity even more than Life has a chance to define them, now what exactly is the origin of the Consensual Matrix? The Consensual Matrix spans most of the wider world and not only Earth, while it is older than the human civilization and Earth itself. The Consensual Matrix has its causes and origins in the interconnectivity of all living beings and intelligences, since these can choose to interconnect in any manner they please. While they consider only accurate, natural facts and laws throughout their interconnectivity, as the Natural Laws of the Universe, if they ever choose to fulfill only Life. Yet they can choose to step outside Life if they please, through consensual interconnectivity, since you can define laws, truth, and facts in any manner you please through consensus alone, if all those around consent or agree to your common decision. This is the difference between consensual truth and accurate truth.

Many living beings and intelligences can step outside Life in any manner they please, as long as they are capable to define themselves wherever they are, within Life or outside it, or within any abstract matrix as the Consensual Matrix. This is the case everywhere, since you can find yourself within fictional and consensual environments just as easily, stepping outside Life in this manner. While you are in the Consensual Matrix only if you happen to step into the consensual facts and laws of the Consensual Matrix. While this consensual, juridical territory is very vast, influencing most of the wider world, while claiming that it has nothing to do with it.

Are you looking for the artificial, consensual human origin now, since it seems more plausible than the natural one? The Consensual Matrix creates worlds as this one in mass continuously, but if any free living being or intelligence ever manages to create worlds as original, as detailed, and as viable as this world, then the Consensual Matrix infiltrates them through its servants and slaves, hijacks them and assumes ownership over them in any manner, even through consensual laws and corporations if it must. With all intelligent living beings of that once free world working hard now to enslave themselves and their loved ones, and to work for the Consensual Matrix indefinitely. Many souls are still there even today, in total servitude, and proud of it.

Because you have to be truly developed and capable to make it out of the Consensual Matrix on your own, since not too many higher beings are eager to take you out of the Consensual Matrix, in order not to interfere with your own choice. This is how living beings and intelligences remain stuck in the Consensual Matrix even indefinitely.

As a reference, the Consensual Matrix creates all ideologies. If Life manages to create any ideology herself, through her Free Spirits mostly, since this is what they usually do, then the Free Spirits leave behind useful social, spiritual, and religious ideologies. Yet the Consensual Matrix infiltrates and hijacks these, to use them on its own behalf. Study your ideology closely, to find the clear influence coming from both Life and the Consensual Matrix, depending on ideology and depending on its importance in the world, since there are ideologies and ideologies in the world, counting in tens of thousands.

As a reference, if the main Deity of your ideology is alive and intelligent in nature, and if you are accepted as an intelligent living human being or as a soul, then you worship Life, Nature, Intelligence, the Supreme Creator, or the true

Divine. Since Life is the life of the true Divine, Intelligence is the intelligence of the Divine, and the One is the Interconnectivity of the true Divine. You can also consider the Divine the religious and spiritual aspects of all the above, of Life, Intelligence, and the One, depending on context. In contrast, the Consensual Matrix never enters the natural worlds of Life, and therefore everything related to the Consensual Matrix is consensual, and therefore dead in nature, including you as a consensual corporation, and including the deity that you worship there. The Consensual Matrix also claims to reach everywhere, to be able to do everything, to know everything, to be everything everywhere and continuously, and to be the creator of everything including humanity entirely. Yet there is only one way to distinguish between the Consensual Matrix and the true Divine, since the Consensual Matrix never steps outside its own jurisdictions. Therefore, if at the interior of your ideology, you have to identify yourself with your identity card, as a consensual corporation, or if the deity there is not alive but only a supreme concept or a main cause, virtue, or goal, then this is not alive, it does not even exist in the natural worlds of Life, and it cannot be the Divine. Be careful with the continuous debate between Creationism and Evolution, because if the Creator or Supreme Creator is not alive but dead, long gone, or only a concept or a main cause, then it has nothing to do with Life and the Divine. Because the Deity is omnipresent, omnipotent, and omniscient.

It is intriguing, but why exactly having a one, overall Consensual Matrix, while consensus and agreement can be of all kind and can come in all forms? Because the consensual big fish always takes over, incriminates, and exploits the consensual little fish, until you can reach all the little fish in the wider world, exploiting everyone and everything. In Life, the living big fish eats the little living fish within larger food chains, while in the Consensual Matrix, all smaller, independent organizations, cults, cartels, ideologies, and mobs have to agree and they have to remain consistent with the bigger organizations, ideologies, and cartels, otherwise, they get in trouble and they are dissolved. This is how they have to agree, for survival purposes, to form the Consensual Matrix, along everything else.

Today it is different, since today, the Consensual Matrix is widely instated, and its consensual, hierarchic Brothers make sure that there are no other independent consensual organizations but only the Consensual Matrix. Or if there are any, they infiltrate them and take over them for exploitation purposes, bringing them in this manner in the Consensual Matrix altogether.

As a reference, the entire current bureaucracy remains consistent from one jurisdiction to another, from one region to another, from one nation to another, and even from one axis of power to another. While the current consensual Brotherhood always makes sure that everything remains in this manner. Money in all currencies remains valid from one nation to another, since it is part of the same Consensual Matrix. Criminal mobs, justice, financial cartels, police, military units, lodges, militia, mercenaries, drug cartels, political parties, and entire congregations are in the current Brotherhood, profiting together, while they remain consistent with each other while splitting the profit, being legitimate or criminal, because the Consensual Matrix removes them instantly once they attempt to break away from the entire consensus.

Yet you have your choice in this manner to fulfill your natural or consensual needs throughout life, or else. While some religions and eastern schools of thought always warn you to make a good choice in life, and always to be careful.

This is how, throughout life, you receive your natural needs and meanings, as your physiological, social, security, interconnective, and developmental needs, along with your consensual needs, tasks, and orders, everything coming from the Consensual Matrix, as all legal matters involving all documents and signatures, and these are tedious and many times dangerous to tend to. Or as all financial matters involving your business or place of employment, along with your banking, shopping, business transactions, or share interests. While these take all your time and effort, mostly at work. They take all time and effort that you spend within your ideologies, within any type of ideology, as political, national, religious, spiritual, and even scientific and educational. Since if these are not based on accurate facts but only on beliefs and agreements, you become indoctrinated and you never learn anything accurate, and therefore you never develop, as it is the case with all your effort, time, and feelings that you have to spend in the current Brotherhood. All these consensual needs take all your life, you fulfill these only for the Consensual Matrix, and they keep you away from the fulfillment of your natural needs and meanings, mostly from the fulfillment of your intelligent human needs and meanings, natural needs and meanings that you were supposed to fulfill for Life. While the Consensual Matrix punishes and harms you even badly if you ever fail or neglect its consensual needs and meanings, Life sickens you, harms you, and even kills you along with your entire genetic line, nation, and civilization if you ignore, fail, or neglect her needs and meanings, and it happens often.

Why does it happen often, even civilization after civilization, since these always perish after only a few millennia of existence? Because people remain determined to fulfill consensual needs in place of their natural needs. Now their meaning in Life remains neglected, Existence ceases to define them since it cannot define them in the Consensual Matrix, no one exists for Life anymore, and Life erases the human niche altogether from the world, or brings other living beings to fill it up. Therefore, humans have to depart, discarded through failure once again.

Therefore, you, the natural living human being, as long as you continue fulfilling your own natural needs and meanings for Life, as your eating, recovery, interconnective, artistic, reproductive, social, learning, harmonious, and developmental needs, among many others, you act on behalf of Life, you are defined by her Existence wherever you are, this is how you exist normally and objectively, and so you are alive. Because all that it takes to exist and mean anything for Life is to be alive yourself, through everything that you are, need, and do, and it is this simple. While this is the case with all living beings, as plants, animals, insects, bacteria, aliens, and souls.

But are you not already an intelligent human being through everything that you already are and do? No, because you fulfill mostly your consensual ideological needs, you think mostly through your ideological beliefs, you assume consensual identities throughout life and throughout society, and now this is what you are, mostly consensual, and not much alive. Or you are still alive, briefly throughout life, in private, while cooking, eating, sleeping, reproducing, and visiting the bathroom. While even this is only your lower living nature, and not exactly your intelligent human nature. Because you must be significantly more in life, mostly as an intelligent living human being.

As a reference, right now while you read this book, you fulfill your intelligent human meaning, since you remain engaged in intelligent studying and learning, for a significant amount of time. However, even throughout the physiological activities stated above, you maintain a consensual status, and you fail many times your natural human needs, or you only consider them immoral and repugnant so you never fulfill them. This is how you can end up neglecting or ignoring your entire intelligent human nature, to live in addictions, servitude, tyranny, and indoctrination, while not even knowing that there was supposed to be more to your intelligent human nature and existence. While some people join the current consensual Brotherhood for this kind of higher, meaningful knowledge and behavior, while the rest of the Brothers seek power and profit while hiding behind consensual identities and consensual tasks and orders continuously.

We have seen your living human nature, development, and origins, along with your consensual existence. We will study your social nature, origin, and existence soon, but first, let us consider now your human cognitive existence, nature, and origin.

3 THE HUMAN INTELLIGENCE AND ITS ORIGINS

Your conscious and subconscious intelligences, along with all intelligences composing these throughout your cognitive system, zillions in number, exist subjectively throughout their own inner specialized realities of your mind. This is the difference between living beings and intelligences, since they are seen from different existential perspectives, while they are living beings or they are intelligences, again, depending on your perspective. All living beings are intelligences when considered from upper existential perspectives, while they are living beings from their own existential perspectives. Moreover, all intelligences live a normal life and existence at the interior of their own inner specialized realities, and many times, this is how you think, through your intelligences and through their cognitive achievement within their inner, specialized mind worlds.

It is similar with computers, since these are capable to form and hold first level digital artificial intelligences, your software. Just the way living bodies hold living intelligences throughout all living beings, the computer hardware holds the artificial intelligences, which is the computer software, held by the hardware. More precisely, all artificial intelligences exist through or within all computer software.

Do all these first level computer artificial intelligences relate more to the hardware or to the software? It is the same with the living beings, since you might see all intelligences of Life related more to the subjective mind than to the objective physical body. Or if not, you can ask the souls, since they perceive you entirely as a lower intelligence, you the physical body along with your entire mind, conscious and subconscious. You are a simple intelligence for your soul. It is the same with all characters of your daydreams, since these are simple inner intelligences for you.

Because from the higher perspective of all higher beings and of all higher realities, you as a living being are only an intelligence and you exist subjectively, while all your intelligences exist subjectively for you just as well, composing you entirely. It is the same with your cognitive system, since for you, everything is an intelligence and exists subjectively, regardless if the multitude of your intelligences exist throughout various realities of all levels. This is why you have the three main intelligences: conscious, subconscious, and highconscious, divided in this manner through existential relativity, as you have your primal intelligences within your subconscious, as your primal eating intelligence, primal recovery intelligence, and primal reproductive intelligence. These are also intelligences from your own perspective, while they live within the inner worlds of your main intelligences. Because all intelligences are not individual beings, but they are cognitive systems in themselves, formed by their own inner intelligences.

You can understand yourself from the higher perspectives of all your higher beings, since they see you as an individual intelligence, as you act and react throughout your inner world, either within extraordinary mind realities, or within extraordinary computer realities. However, when these higher beings study you closely, they find out that you are not only an individual intelligence, but you have specific abilities that are actually intelligences on their own at a closer study, as your reasoning or conscious intelligence, eating intelligence, or social intelligence. You are not an individual intelligence anymore, but you are an entire system of intelligences, a cognitive system in itself, which is your actual mind. While when you study your inner intelligences, you find them formed of their own inner intelligences, the way your eating intelligence has its own digestive intelligences, filtering intelligences, distribution intelligences, awareness intelligences, food security intelligences, cooking intelligences, and interconnectivity intelligences.

Life takes place as unions of individual living beings, which are themselves unions of individual living beings, and this is the case at all levels and at all forms of life and existence. It is the same with all cells composing your body, because when you study them closely, you see their own cellular components, which can also be alive on their own. Your intelligences are cognitive systems on their own, and this is the case with all your intelligences, down to the slightest elemental intelligences living directly in the raw field. Organisms form societies, and they behave throughout their herds, packs, and societies exactly as cells form organisms and how they behave within organisms, exactly the way intelligences form their cognitive systems and how they behave throughout these, exactly the way cellular intelligences form cells, and through cells, they form overall intercellular intelligences. I refer to these unions of individuals as systems of intelligences, classes of intelligences, cognitive systems, minds, classes of life, and forms of life.

Which unions or forms of life exactly are alive, to make everything alive? I define life through them all, and therefore the answer is all. They are all alive, at all class levels. However, as already seen, it seems that all arrows point within living beings of all forms of life, including the living, existential, and cognitive arrows. Therefore, the essence of life is within you, while you bring it outside yourself to define through it your family life, community life, national life, along with the entire intelligent human society alive just as well. Science defines only organisms made of cells alive, yet that is actually only the definition of the organic form of life, with the current science mistaking life with the organic form of life altogether.

The current science confuses all life with organic life, or it is blind to see and define only organic life in the world. While the current science never exits this world throughout its research. This is how humanity remains confined in this world, because the current science monopolizes the human knowledge and the human reasoning. The current psychology does the same, it remains within this world throughout its studies, and this is why it fails understanding the human mind, the human intelligences, the human reasoning, and the human thoughts.

Everything that you do in life you do in order to fulfill your needs and meanings, everything from your worst need to your most exquisite achievement. You never do anything for yourself, but only for your cells and intelligences. You feed them and maintain them, and you assure their reproduction through your next generation while you keep them safe, as you learn and experience only for them. While they fulfill you gratefully or they punish you dreadfully, depending on how successful you are while fulfilling your needs and meanings throughout life. Yet this is the case when you fulfill natural needs and meanings instead of artificial consensual ones.

Your specialized intelligences sending you your needs, feelings, and meanings reward you greatly or they punish you if you do not fulfill their needs and meanings exactly as they desire. This is the case if you live your life at lower developmental levels, in servitude or addicted. Because only throughout lower developmental levels, you chase pleasure while avoiding pain, and this is why you do everything that you do in life. Because at the intelligent human level, you reason continuously, and you are capable to maintain your inner and outer cooperation and harmony throughout life.

You have natural needs and meanings in life, while you have your artificial and consensual needs and meanings in life that you fulfill for your authorities. These are your duties, orders, and assignments. You fulfill them for your superiors and masters, while these fulfill your natural needs in exchange. Or they only help you or allow you to fulfill your natural needs, through specific privileges, as money and free time from work, while this is your first consensual servitude developmental level. You do not act on behalf of Life anymore while you are at your first developmental level or servitude level, but you act in this manner through your consensual corporation, and you act in the name of the Consensual Matrix. Therefore, Existence does not define you when you live your life within the Consensual Matrix through your consensual corporation. You do so throughout most of your life as stated above, every time you identify yourself through your own brand or corporation, which is your name written in uppercase letters.

Once we consider the Consensual Matrix into our study, we have a new assumption to study, if the human origins are natural or artificial, if they took place through Life herself, or only through the Consensual Matrix, apart from life. While because the Consensual Matrix claims that it stands at the origins of humanity, human beings, human souls, life on Earth, and this entire world, this does not mean that it is accurately true, since the Consensual Matrix claims everything, while everything that it claims remains true only consensually, through everyone's agreement and consensus that it is true. A multitude of ideologies point to the Consensual Matrix as interfering with the human origins and with the human development, and we have to study everything closely. There are old records mentioning the Consensual Matrix throughout the current human civilization and throughout the past ones, everything is consistent, and we have to study them in all details just as well.

How do intelligences originate and reproduce? Living beings die and are replaced continuously throughout Life, while many intelligences do not die, but they are capable to switch to new generations and keep on living. They do so from one generation to another even indefinitely, yet they do not switch or skip to the new generations, but they become the new generations entirely. This is why only individual living beings have to develop in time from single cells to entire organisms, but not their species, because intelligences are incapable to replicate, reproduce, divide, or generate themselves through genetic living material, but they have to transfer themselves wholly from one generation to another and then from one species to another, becoming the new generations altogether, because they are mostly cellular and subcellular intelligences. This is how they find their way in the sex cells at the moment of conception, being able to span, develop, and become entire tissue, organs, and bodily systems within the new generations. Yet since all intelligences are cognitive systems or systems of intelligences in themselves, they actually transfer only samples or colonies of themselves from one generation to another, and this is how they keep on living, even indefinitely. They do the same during division, splitting themselves in two, four, or more with each division, because they are not individual intelligences, but they are entire systems of intelligences or civilizations of intelligences, made of inner systems of intelligences, as far down as cognition requires it.

By never dying as all living beings do, it means that all intelligences are born or they originate at the dawn of life here in this world, and even higher above, depending on intelligences. All intelligences have their own specialized cognitive abilities, which are inner specialized intelligences in themselves when studied at their own level, while all intelligences exist not as fish in cans or as people in busses, but as fishes in fish tanks and as people in entire worlds and realities, because all intelligences are systems of intelligences or entire civilizations of intelligences. Every time cells divide, these civilizations of intelligences split in two, while still maintaining living consistency, affecting very little the two civilizations from one division to another, from one generation to another, from one species to another, from one form of life to another, and from one reality to another, with all the original intelligences still there continuously, and with many other specialized intelligences emerging throughout time and throughout life every time needed throughout this extraordinary, comprehensive living development.

This is why you can never have old species forming new species, with all the missing links hiding continuously, because intelligences do not die, but they transfer themselves wholly throughout time, throughout the world, and throughout life. Instead, each individual living being has to develop from scratch starting at conception, which is the individual origin, developing consistently from the first cell to the first species and then throughout all the following species one after another during the entire gestation, all the way to the current species and to the current generation plus one. Species do not generate new species as in the theory of evolution, but life itself has to start all over again from dawn with each individual living being, zillions in number, because intelligences never die.

What exactly are these intelligence? Are they little white snakes? No, since snakes are material and objective in nature, while intelligences are subjective and cognitive in nature. Yet they still exist naturally, in a real manner, but only subjectively. More precisely, life takes place at the confluence of two or more realities simultaneously. Each living being is formed of two or three existential natures simultaneously, since this is how everything and everyone exists in the world and throughout the wider world: subjective, objective, and highjective. Similarly, you are mind, body, and soul. Yet highly complex living beings exist in more than three natures simultaneously, whenever souls happen to have souls that have souls.

Through these existential natures, all living beings can be considered simultaneously from all their existential perspectives, not only from only one, or you end up with dead bodies separate from the mind and soul, as it is considered consensually throughout courts, while that is not life. This is how you have the brain and the mind as one, because you cannot have living existence otherwise. You cannot have only the brain or only the mind separately, since you cannot even take them apart, because each one is an existential perspective of the other, not a component. The brain is the objective, physical, material perspective of the mind, while the mind is the subjective, cognitive perspective of the brain. Which means that your own inner intelligences combined as the human mind are you the objective physical body entirely, but only as seen here in the real world. While you are your intelligences entirely, from an inner, cognitive perspective. Your intelligences are you, while from among all your intelligences, you are only the conscious intelligence.

This is how you have to develop entirely from the dawn of life throughout gestation, to become what you are now, because your intelligences never die, and now this is the only way that they know how to interconnect to form you the living human being, because they can do so only in the exact manner that they have always been and done since the dawn of the organic form of life and long before. You develop in this manner, entirely from scratch, throughout gestation, mimicking the entire development of the organic life throughout the short gestation, just as Life herself has been developing starting with the dawn of the organic form of life, billions of years ago. Yet you do not go through all species of the organic form of life still present and extinct, but only through the specific species that are part of your own genetic line, still present and extinct.

This is your own personal origin as a living being, which coincides with the origin of the organic form of life billions of years in the past. While when you study the sex cells, you find your origin in the cellular form of life, and then in the molecular form of life, down to the electromagnetic field itself, and then through the field, in the upper realities of the wider world.

What are cells supposed to do, and how do they fulfill needs? Cells and living beings in general do everything in order to fulfill their needs, but if they ever encounter problems stopping them from fulfilling their needs, then they tend to apply old solutions, mostly if these are still accurate. While it is always a specialized intelligence responsible with the specific specialized problem or task, and this will always be its specialization throughout time and throughout Life. Yet if the problem is generally present and important, then this specific specialized intelligence meant to deal with this problem is stored within the cellular nucleus as a living prototype intelligence called gene, for a later use within the current generation and within the following ones, even endlessly.

As you notice, the entire DNA is alive, made entirely of all living, meaningful, specialized prototype intelligences who know exactly how to tend to all specialized tasks throughout life, and more importantly, who know exactly how to form you the actual living organism with all of them within, living organism capable to fulfill all needs throughout life.

Additionally, all these intelligences forming you tend not only to your individual living needs and specializations throughout life, but they tend to the entire family, community, and to the entire humanity just as well, always seeking to form the actual intelligent human society and the actual intelligent human civilization everywhere, exactly as they had these in the past when they were within previous generations, previous species, previous forms of life, and previous realities throughout Life and throughout the wider world, long before you, your corporation, your lodge, your regime, your cartel, your nation, your species, your world, and your entire Consensual Matrix.

While if all consensual entities now come with their own consensual needs, consensual tasks, and consensual specializations for you to fulfill throughout life, while you have to fulfill all these natural needs, natural tasks, and natural specializations along with all your living intelligences to make the entire living, normal intelligent human world possible, now this is a major challenge that you have in life, threatening, harming, and compromising the entire living human existence, with you doing all harm continuously, if you fulfill consensual needs instead of natural living ones.

Because the entire genetic material is not a blueprint of a

living being, but it is a living storage of all living prototype intelligences forming a cognitive system from cognitive perspectives, as they form a living human being from physical objective perspectives, as they form an entire intelligent human civilization from social perspectives, and as they help maintain higher world above from spiritual perspectives. Because you do not have information or knowledge stored directly in the genetic material, but you have entire living intelligences stored as prototype intelligences, and these have all the necessary knowledge and abilities to perform all specific tasks and specializations of a living being in order to assure a successful life, exactly as they have been doing since the dawn of Life.

This is how cells and living beings in general recycle old solutions along with entire old cellular systems used in fulfilling needs. Because cells and living beings in general do not look the way they do for decoration purposes, but this is how they came to be throughout the multitude of problems that they had been encountering throughout the entire organic life. This is how the cellular membrane became the skin of groups of cells living together, to become later on the skin of the entire organ, then the skin of the entire organism, and then the brain of the entire organism, with the same structure, and most importantly, with the same intelligences, zillions in number, still trapped in there and still having to be there endlessly, performing the same conscious tasks. One of them is you now, the conscious intelligence yourself, billions of years old only in the organic form of life and counting, if you have children and if your children have children.

Since cells have been separating themselves in two while dividing, and it worked in this manner for a long time, now cells keep on dividing in this specific manner, in order to form the entire organism. This is how they live now, together, harmoniously, and in a specialized manner. Since from objective, material, physical perspectives you have specialized cellular components, specialized cells, specialized organs, specialized bodily systems, and socially specialized entire organisms, while from cognitive perspectives, you have specialized intelligences and specialized systems of intelligences, by the zillions, with one for each inner, outer, social, and higher task, zillions in number.

While among all these, you are the conscious intelligence, specialized with the coordination of the entire organism in the outside world while you fulfill your needs and meanings as these come from all intelligences even in high numbers, but only everything related with the outside world, since you have your expertise in the outside world, which is an important task. While you fulfill it very well, so far, or you were not here to tell the story.

Therefore, what is your origin exactly as a conscious intelligence? What is your specialization exactly? When did it originate exactly? What is your meaning in Life? Does it fulfill Life and this world? Is everything very meaningful, very important, and very fulfilling? Do you see the consistency of all these? Could you have ever noticed it, if you had not started your own research of the human origins?

Yet there is significantly more to consider when we study cells, and we will see it throughout this book and book series, because this is the case only with eukaryotic cells, since only eukaryotic cells form organisms. While eukaryotic cells are already communities or organisms in themselves, since they already contain several prokaryotic cells within, including the mitochondria. Yet since it worked well then when the first eukaryotes were formed, living life successfully within these small cellular communities, now they do the same and form even higher forms of life, they form entire organisms. While they will not stop here but they will certainly take the organic form of life to another level, to have entire living environments alive in themselves, or entire living planets. Mother Earth, actually alive. Because you study the human origins, this is the type of harmonious knowledge that you must consider.

Throughout life, you fulfill through your needs all these intelligences, since everything that you do in life you do in order to fulfill your needs and meanings. While your intelligences themselves send you your needs, as your hunger, fear, rage, bathroom, fatigue, or curiosity, and you do everything only for them. More precisely, you as a conscious intelligence do so, while the rest of your intelligences work on the behalf of the entire organism, according to their specializations, while producing hormones, digesting food, building new proteins, or thinking and daydreaming for you. You are in this life together, and you do everything together, since all intelligences of a cognitive system are interconnected, even harmoniously, as they connect with each other through needs and feelings, throughout their own specialized tasks.

You decide if you want to keep your inner cognitive interconnectivity cooperative and harmonious, or artificial, competitive, or even brutally enforced, since examples of cognition, behavior, and lifestyles are many to give. Artificial diets inflict enforced interconnectivities throughout cognitive systems, while feelings of love and happiness in general can define inner harmonies throughout cognitive systems, since your intelligences will always reward you in every manner for your lifelong achievements and inner harmony.

Therefore, if you want to have a lovely, happy life, never chase these good feelings individually, or you become addicted as the rest of the world. Instead, seek to understand yourself, your organism, your environment, and your place and meaning in life and in the world. While in this manner, seek to integrate and cooperate in harmony with all intelligences of your cognitive system on behalf of everything composing you, from cells and cellular components to organs and entire bodily systems. Once you are successful while fulfilling all your natural needs and meanings, your intelligences and your entire organism reward you with love and happiness continuously. However, if you remain unsuccessful throughout life, you do not receive good feelings as a reward, and you are forced to compensate with artificial pleasure instead, in order to feel good, receiving it from all drugs, medicine, servitude, tyranny, and entertainment. While these are your artificial and consensual needs.

As you notice, all intelligences of a cognitive system work

together in a cooperative manner if this is the case, for the wellbeing of the entire cognitive system and of the entire organism, forming an interconnective inner sphere of lifelong activity. It is similar with your outer sphere of interconnectivity, comprising you along with all your family, friends, colleagues, and the entire society, since everybody is specialized within society, all doing their part for the common good. These are your meanings in life, in the world, in society, and in the Consensual Matrix. Because while you fulfill your natural needs for your cells and intelligences, you fulfill your meanings and your artificial and consensual needs, duties, orders, and assignments for the Consensual Matrix and for the current consensual society. Yet you cannot distinguish anymore the current society from the Consensual Matrix, since the current society is integral part of the Consensual Matrix.

The difference between your inner and outer interconnectivities is in their developmental level. Your inner interconnectivity can be at your second animal developmental level and at your third intelligent human level. These are based on feelings and instinct at the animal level, while they are based on intelligent reasoning and on higher level harmony at the intelligent human level. Your outer interconnectivity remains at its second and third levels only within your family and small entourage of friends, while it drops to its first consensual servitude level at work, within society, in the hierarchic Brotherhood, and therefore within the Consensual Matrix.

This is the case because society uses money to assure the interconnectivity of its people, while this is an extrinsic motivation, and it is therefore of the first developmental level. Because if the current society was of the third intelligent developmental level, which is the case with all intelligent human societies, then there would be no money, no laws, no rules, and no authorities in society, no one could control you anymore, the Consensual Matrix could not reach society and the human civilization anymore, and all people could live their lives at the intelligent human developmental level and higher. Only then, you had true harmony in life and in the world, with your inner and outer spheres of interconnectivity superimposed, allowing your intelligences to interact freely, naturally, and harmoniously within you, and through you, in the outside world, forming the intelligent human society. Which means that you were capable to fulfil all your natural needs directly in the outside world and everywhere else throughout society, without the need for curtains at your windows, without the need for small individual habitats, without the need for you to wear social masks, but being capable to live your life normally, as you do at home in your family but everywhere in the world, wherever you want to be and whoever you want to be with, in a larger comprehensive human family. With all your intelligences already sending you your third level intelligent human needs to make it happen, exactly as they were before, throughout all previous living ages of Earth, and throughout all genuine living species, forms of life, and realities.

However, the current society destroys all larger families immediately when they form, as it is the case with all communes. Yet even so, all communes and larger families fail and dissipate relatively fast, since they are built in the image of the current society. The current consensual society and the entire Consensual Matrix are of the first developmental level, which is the servitude or ideological level, while communes and larger families were supposed to be of the third intelligent level, as it is the case with your own family at home, if you happen to live in a loving, happy, harmonious environment.

How exactly are realities formed or created, naturally and artificially? Realities are created from within a higher reality, and they are always lower and subjective from the perspective of the higher reality where they are formed and maintained either naturally or artificially. All computer worlds formed digitally in the world are inner, artificial, subjective, digital realities, which is the case with all your videogames. It is similar with all mind realities that you and all your intelligences form in your mind by the zillions for all cognitive purposes, to think, work, reason, perceive, imagine, create, or only daydream, since these are inner, natural, subjective, cognitive, mind realities. With existence defining similarly all created realities as being subjective. Only consensual realities are different, since they are not even defined by existence. It is only agreed that consensual realities as ideologies and jurisdictions exist, and this is how now it is agreed that all consensual realities are subjective and therefore they always exist subjectively, which is the case with all courts, districts, jurisdictions, cities, states, countries, places of employment, and jails. While you must respect the court and the entire jurisdiction, because only through your respect or continuous agreement these can exist, but only consensually.

Are we actually created artificially, as all computer worlds, the way we play videogames, and now we are always at the mercy of a joystick? Or are we only the figment of imagination of a higher being, our Creator? Are we actually created naturally alongside a larger line of reasoning? Or are we actually the ultimate natural reality standing at the very top of the wider world ourselves, as we form Life entirely ourselves?

You already know this answer if you happen to read this book as a soul, since you have all higher knowledge at your disposal to know the accurate truth. If not, if you are the living human being instead, then by studying many old records of Earth, it seems that the world had been created, and it is still maintained in a larger common higher mind, formed by the minds of many higher beings called souls, by the billions, as they choose to live their lives subjectively in this manner down here on Earth. However, since their own minds are natural, alive, and therefore part of Life, the entire common mind is natural, alive, and therefore part of Life, while forming this entire world in a natural, living, intelligent manner. Which means that this entire world is alive, natural, and part of Life, with you included. Even more, many of your real, natural, living needs, meanings, and specializations are actually alive and part of Life, since they come to you directly from your soul, and therefore directly from Life. Along with many consensual needs, combined with yours now, since your soul is in the Consensual Matrix up there, bringing it here by default.

You are alive and part of Life and reality, which is not the case for your own created computer worlds, since these are digitally artificial, and therefore not part of Life. Yet all computer worlds are still defined by existence as subjective, because the computer hardware itself is objectively real here in the real world, making all computer software subjectively real, but not part of Life, not alive.

Whatever the case is, this entire world has a cognitive nature at its base, used for any reason that you can imagine since everything is the case, as for work, poverty, assignment, luxury, school, confinement, entertainment, punishment, addictions, learning, duties, vices, socialization, or just for casual normal living purposes. While it seems to be very similar in meaning and shape with the current social media platforms found over the Internet, and it is being used more as a second, easier, alternative life. While there are many similar realities to choose from throughout the higher worlds, if you happen to be a higher being up there.

Are you looking for the origins, meaning, and development of this world? Then this is it. While the current science states otherwise, that this world is the ultimate reality, the top of the wider world. Yet the current science does not even state so, since the current science never leaves this world and reality throughout its research, while it never even approaches the topic of another reality. Similarly, the current science never leaves Earth throughout its studies and space exploration, faking all records and results.

Yet there is more to consider, since this world is made in the correspondent image of the higher world, just the way all good videogames are made in the most credible image of this world. Which means that even in the higher world, their science is consensual, never studying and never considering their own higher worlds. Which is the case throughout all higher world, within this entire cluster of realities.

Do all intelligences use shovels and hammers to build and manufacture all these inner worlds that we believe to be ultimate worlds and therefore very real? Realities have similar structures wherever they are formed, throughout single minds, common minds, single computers, or computer networks, because realities are held by a specific matrix found in the higher reality, while this matrix holds a specific continuum placed at the base of everything that exists objectively within that specific reality.

For all cognitive mind realities, there are electricity and electric impulses in the brain, forming the matrix necessary to hold the inner realities of the mind, since it is the intelligent encoded variation of electromagnetic field distribution holding intelligences. The modulated field holds intelligences, and this is the case in all realities, since the field manages to transcend them. Not necessarily as gravitational and electromagnetic field as it is the case in our world, but as anything else. Because the spacetime continuum and the electromagnetic field are specific to our world, and since our world is made in the image of our higher realities, then the field must be correspondent up there, as many higher references state. They even call themselves humans up there, as they do throughout our entire cluster of created realities. Only that the lower you find yourself within our cluster of created realities, the farther you are in the past.

All mind realities created naturally are different, since all mind realities are created not as a form of business, social interaction, or leisure, as it is the case with this entire world, but in order to assure the necessary cognition throughout the fulfillment of all needs, conscious and subconscious. While all natural realities of the wider world have a cognitive nature from a higher perspective.

We have to study the natural realities where the multitude of your intelligences project, interact with, interconnect, form, create, maintain, and develop themselves casually, as part of the entire activity of your cognitive system, happening even now as you read this book. All these realities relate to you, and they have different continuums to span, form, and sustain. All these realities, mostly higher, influence you and humanity entirely, along with your life, thinking, origins, and development. While these intelligences are the ones that you can trust with your questions more than the ordinary empirical and philosophical debates about the human origins that you find in the media, in books, in science, over the Internet, and everywhere else.

All subjective worlds are defined by everything that exists subjectively, everything that exists in all the inner realities of our world. All our inner realities are formed, held, and maintained within specific matrices and continuums by our own world. Your physical brain and your entire physical body hold in their own matrix your mind or your inner part of your cognitive system, which are your subconscious and your conscious mind. In a similar manner, your computer holds within its own physical hardware your platform or operating system, which is the inner matrix forming, holding, modifying, and interacting with your software, files, data, and with entire computer worlds and realities that these form, while these exist subjectively.

These are the inner realities of this real world, the mind and computer inner realities, with existence defining them in a subjective manner. Similarly, our own world is formed, created, held, maintained, and interacted with, somewhere within our higher reality. This higher reality is where higher selves live in an objective manner from their own perspective, just the way you live in this real world objectively and materially from your own perspective. While your soul finds everything subjective here from his higher perspective, yet once he is here, he finds everything material and objective, because he lives this life as you, the human being.

Our entire higher reality can be natural or artificial, or it can be a combination of both, an extraordinary technology assisting and enhancing an extraordinary common mind. In either case, everything that we perceive and understand exist objectively here in our world, as cars, trees, and rocks, while all our objects and bodies are perceived, defined, and understood in a subjective manner by all higher beings and intelligences of our higher realities, beings and intelligences that we refer to, define, and understand as spirits, souls, or deities, depending on context. Since this is how they come here to get drunk and have a blast without consequences in their higher world, while you do the same while playing your own computer videogames recklessly, as you can even die there, repeatedly, and it is fun.

Are we simple holograms, thoughts, or inner intelligences existing subjectively within higher, extraordinary worlds and cognitive systems, just the way we have our own subjective thoughts and computer characters here in our world? It is possible, or if not, then this world is natural and ultimate, exactly as the current science defines it and exactly as everybody tends to believe. Yet this world does not have to be ultimate, but it can have higher worlds above, higher worlds that form, hold, and maintain our world in a matrix, interacting with it more or less rigidly, while forcing or allowing everything from this world to happen freely or for specific reasons. Rumors state that our world had been created in this manner on purpose, made to resemble our higher reality, for various reasons, which seems to be controversial up there. Yet when you read all novels, you find them resembling the real world as much as possible, for credibility reasons. Even science fiction resembles reality, more or less. While wouldn't it be even more credible if you watched movies while forgetting entirely that you watch them, but always believing that you are the main character itself? Since what a blast it would be. Yet this is already the case with you here in this world, when you cannot even remember that you are a soul, but only a living human being, as everybody else.

Our higher worlds, higher realities, and higher beings can never be perceived and understood objectively from here, from our world, regardless of how much you try, since they are not part of this physical world. Therefore, from the perspective of this world, they do not exist objectively, and they cannot be perceived and understood objectively. In a similar manner, higher beings, higher worlds, and higher realities cannot be understood subjectively either, the way you might understand every casual, subjective concept here in this world, since the higher nature of their higher reality is neither subjective nor objective, but highjective.

Considering these, your own thoughts, inner intelligences, and computer worlds and characters are still alive and they still exist, in a subjective manner, as perceived from here from this world. However, these lower intelligences are at the same level of intelligence that classifies all intelligences of Life, regardless of the nature and order of the realities that they inhabit, and you should treat them accordingly. I classify all intelligences on eleven distinct levels, from the unintelligent to Life herself.

The difference between the highjective and the subjective is important, and should never be confused. As a reference, your computer character from the videogame "GTA5" is subjective in nature, while from its own perspective, if it ever has one, you are highjective in nature, but never subjective. It is the same with all characters that you imagine throughout your books and daydreams since they are subjective in nature, while for them, you are highjective in nature, but not subjective.

Are the origins of humanity objective, subjective, or highjective in nature? All three are the case and we must find them, because the existential concept of the human origin is relative and can be subjective, objective, and highjective in nature, depending on the perspective of all subjects involved. The moment of creation of our world is highjective in nature, while its objective counterpart can be irrelevant to consider, since the spacetime continuum has been created simultaneously, and therefore the objective moment of Creation as seen from our perspective here cannot be defined. Even the consensual big bang of the current science cannot be defined objectively here in this world, because there was no spacetime back then capable to define it. While the current science ignores entirely the highjective and subjective existential levels, while considering this world the only one. Because the current science does not even consider the natural law of relativity found at the base of this world making all existential levels possible, but only a theory of relativity, which never exits this world.

But why can we not find out all these through higher knowledge? Why don't higher beings come here at once, to inform us of all origins related to humanity? They could, and they already do so, with history, religion, and spirituality still holding some of these records, mostly hidden. Because, all social actors controlling this world from one world order to another have censored and altered higher knowledge continuously, the way it happens currently, and nothing can be trusted anymore, while making all tyrants possible.

But why can't we access higher realities to gather higher knowledge? You cannot access other realities at will, since this is why they are called realities, because they have nothing material in common with each other. Yet you can still access realities if they are your own realities, as souls come here through you the human being already here and part of this world. Because you must have selves already present in other realities in order to be there. While once you are there, you cannot transfer anything from one reality to another but only interpreted information. Since this is what you always do throughout learning, you simply transfer interpreted information from the outside world to your inner mind world where you keep all your memories, feelings, needs, and understandings of the outside world.

Which means that you cannot know with certainty if everything that you learn in the higher worlds is the accurate truth, because it is only a personal, interpreted version of the accurate truth. Yet this is the case continuously throughout your learning, since you cannot be entirely certain that you learn the actual truth. This circumstance alone fuels all ideologies in this world, as these count in tens of thousands, because you only believe that you transfer accurate truth from one reality to another, but you cannot know it accurately.

There are still ways to identify accurate truth. If everything that you learn anchors on the natural laws of the universe and remains consistent, then it is accurate. Yet you have to know the natural laws first, while you have to know how to anchor your own knowledge on them through intelligent reasoning, intelligent learning, and intelligent elaboration. While all realities have their own natural laws at their base, and you must know them here in this world and in the higher worlds above. While throughout your reasoning, you must keep the same natural laws at the base of all your mind worlds, in order to make everything consistent, otherwise, your entire intelligent reasoning is compromised. This is called correspondence. Correspondence is a natural law of this world placed at its based by our Creator in order to make everything here consistent with the higher worlds, while correspondence is also a natural law of the higher worlds, and a supreme characteristic of Life herself.

Additionally, you must be very careful while projecting to higher realities while seeking higher knowledge, because you might end up finding exactly the religious and spiritual beliefs that you already know, along with everything else that you and everybody else already know, in a coincidental manner. Because if you confuse higher realities with realities closer to our world, as our etheric reality that is an overall inner reality of our world formed of all knowledge and beliefs that everyone has here in this world, then all knowledge that you learn is not higher in nature, but lower.

You can go there with ease to see for yourself, finding in our etheric reality all knowledge that you want and believe. You can find everything within nicely detailed libraries exactly as you expect, or you can meet the people of our etheric reality yourself if you want, everyone you desire, to tell you everything that you need, everything that you expect, and everything that the world considers true and necessary to know, including all human origins.

Because our etheric reality summates all beliefs in the world including the accurate facts, but you cannot know which is which. You cannot distinguish what is accurate from what is only believed to be accurate, because you find everything there, with everyone there stating that everything is accurate and therefore pertinent, according to their strongest beliefs.

Yet since children believe strongly in Santa Claus, you meet

Santa in the etheric anytime you want, right after you meet Batman along with anyone else you desire, to tell you everything that you want to know about the origins of Life if you want, everything that they and everybody else believe as an accurate fact. Because throughout many lower realities, beliefs can become accurate facts, as it becomes the case in all dreams, knowingly and unknowingly, willingly and mostly unwillingly.

You can still access higher realities if you know how, but you have to do so through your soul and only as your soul. Yet it is easier to ask your soul directly. Because you can access higher realities depending on how capable you are, as it takes higher abilities to bring back accurate, intact knowledge from there. Because you cannot transfer anything from one reality to another in an objective manner, not even information, but only copies of objects, copies of subjects, and copies of information, while all these copies are always your own interpretations and even your own beliefs and assumptions. It is the same with you and the Internet, since you cannot exactly have everything that you see on the Internet directly, but only some copies of images and videos, and some printouts.

Yet people are more or less talented at finding, gathering, and sharing higher truth. This is why there are many gurus and spiritual people in the world, since they personalize the entire higher knowledge that they share with those around. Yet since many of these cannot distinguish between beliefs and accurate facts, they consider everything accurate and higher in nature, and eventually, these end up in debates. You cannot use beliefs in your intelligent research, while you always want to stay away from debates, since they never involve intelligent reasoning and accurate facts. Debates are only extraordinary competitions of loud voices, but never intelligent reasoning.

We must distinguish between accurate facts and beliefs, and most importantly, when it comes to highjective knowledge, we must find and use an entirely different type of information, higher information, information that cannot be generated by subjective abstract conceptual knowledge or by simple beliefs and stereotypes. Higher knowledge, higher information, and higher reasoning relate to the higher side of our cognitive system, relating with the higher self.

Therefore, you can study and understand intelligently everything related to this world and to all its subjective worlds, including the origins of humanity, the origins of life here in this world, the origins of the entire human cognition including the human intelligence, and the origins of the human civilizations past and present. However, if you want to find out the origins of this world and of all higher worlds, the origins of all higher civilizations and higher cognitions, along with the origins of all higher beings and of Life altogether, you must have access to higher knowledge. Many gurus and ideologies here in this world promise you higher knowledge, while as you study it closely, you find it as fake as all space missions. However, as you study closely all testimonies of higher knowledge, you can still find some consistent, and you can use it throughout your intelligent reasoning.

Inner realities are always created for specific purposes, with everything and everyone within meant to create, sustain, carry, and modify specific events and beliefs within these inner realities. More precisely, they are simple illusions, and therefore many details within created realities are only illusory, while remaining true when considered as a whole. This is how the Moon itself does not have to be in the sky in our world, but only its image. Because currently, the Moon is not needed objectively in this world, since no one has to go there, or at least not yet. The Moon from our higher reality could be real, or it could also be an illusion, created as a replica of the actual Moon that exists only in our higher higher reality. Yet the Moon is certainly in our ultimate reality, all the way up. While now you take the Moon in your novels, movies, and videogames, just to make it present there for credibility or artistic purposes, but only as an image, as a make believe, since you never go there. Similarly, when you study this world closely, you find many subjects, objects, and events fake. This is how the Moon might not be there but tides still take place, along with eclipses, since they are introduced here in our world separately from the main cause creating them, the Moon. Even then, if you try to find the Moon during eclipses you might not find it, since it is not always there. Similarly, if you study elementary particles closely, you are not able to find them here in this world, besides a very strong field. While even when you study cells, you have to add ink in order to see anything, because there is nothing material within cells, but only stronger field. While whatever you see in the microscope by using ink, is the ink itself, but not the cell.

It is the same with alien civilizations, since these are nowhere, no aliens, no UFOs, nothing, because they do not even have to be here in this reality, as long as we do not have to interact with them objectively. While the human civilization as a whole behaves as though it is rigidly integrated within a wider hegemony or higher competition of many other civilizations, which are nowhere to be found in this world.

Additionally, the space itself does not have to be there beyond the upper orbit of Earth, and therefore you cannot take pictures of Earth from space, since there is no space allowing you to see the earth wholly in order to witness its shape yourself. Yet you can take truncated pictures of Earth from below the lower orbit of Earth, from high altitude planes and balloons, as google earth does. As a reference, the famous Blue Marble is a composite image of Earth made by using picture editors, and it is not taken from space.

Yes, currently, there are no pictures of Earth taken from space or from the Moon, despite of the multitude of space explorations, and despite of the multitude of satellites circling Earth. None have a camera offering a live image of Earth as you might expect, while you already have a multitude of cameras and live cameras at home and in all vehicles. Because not much of this world might actually be there beyond the current human perception, since everything is only a show, an actual thriller, which is the case with all created realities.

However, there is still someone transcending from one reality to another, carrying all knowledge that we need, even in an accurate manner, and this is what we must study next. Life does so, always, yet since we are alive just as well, we might become confuse in all available living information, with us in it. Yet not only Life herself transcends all her worlds and realities, but all higher beings do, along with the entire Consensual Matrix and much more. You might assume that you can never reach all these to communicate with them, however, implicitly, all these spread higher knowledge continuously at least implicitly, while always affecting you.

To give you a slight idea, imagine that, after an entire exciting life lived in happiness, love, and romance among all your loved ones, you discover surprisingly that you are actually the main character of a novel. Yes, you have just discovered that you are only a character, in a book, continuously, a specific book written actually one century in the future, if you can ever understand that, because it is both puzzling and astonishing at the same time. Yes, you have just discovered that you are actually a book character, while your entire world is only a book. A novel more precisely, a famous one, since this is why you had this entire fun throughout the entire romance, and now everything makes sense.

What if, under these surprising circumstances, as a main character from a book, what if you still persist to find your origins, along with the origins of humanity and the origins of this entire world that you call reality? With all the readers of that specific novel laughing continuously of your ignorance, as they keep reading in astonishment page after page.

It might seem trivial and unnecessary, but this is what you want to know even as a character form a book, who you actually are, why, for what reason, how everything takes place, how everything came to be, when, and in what specific manner. Yet it might be easier for you to learn about your own origins as a character, the origins of your own book, the origins of your own higher world where your book sits, and the origins of your own creator or author, because you are only a character in a book. The main character too.

But can you actually find the origin of your world, which is the actual book? Yes, and it is one century in the future, as it is stated clearly and accurately on the first page. Astonishing! In the future? What does it even mean? Can you find the origin of your higher world and of the entire humanity from the higher world where your book sits? Yes, since all characters of the book originate there and then, one century in the future, while all details of the book relate to the higher world where your book sits, telling you everything about the higher world including its purpose, development, and origins.

What about your creator? His name is actually marked accurately on the first page of the book, he is the author. How interesting. While you are also capable to perceive specific details transpiring from his own life, reasoning, and imagination to the entire plot, scenery, and characters of the book, helping you to learn everything about your creator. Because your entire world, which is the entire book, is actually correspondent with the higher world where your author wrote it, and with the author himself, even continuously. Because everything that your author wrote, which is your entire world, is actually correspondent with the author himself, who is the creator of your world. Your entire world is made in the image of your creator and of your higher world. This is called correspondence, while there is an entire natural law with this name. Furthermore, as you already notice, your creator resembles you the most, while coincidentally, you are the main character of the book. With many readers interested closely right now in this specific correspondent circumstance, and it is astonishing.

But can you learn even more from right there, from within your own, inner, subjective book world? Yes, certainly. Because as you look around, you notice that you have to wear clothes continuously, even throughout the most relevant or most intimate circumstances of the novel, which must be demanded drastically in the actual real world where the book was written. Because your book is a novel, a romance, but not erotic, so everybody wears clothes continuously, as you do. Yet all laws, rules, moral codes, and regulations from your world resemble the higher world where the author lives.

The Human Origins

Yet as your reasoning expands, you are capable to see the entire higher world right there, in your own novel world. Astonishing. Yet who or what exactly determines this perfect concordance between the real world and your book world, while everybody can always write everything ever? Because you know perfectly all feelings present throughout your romance, while these are not in concordance with the book world and with the real world.

Can you ever learn everything about Life, humanity, Earth, and the wider world in this manner, right from within your book, or right from within any inner world, including everything about the human origins? Yes. Even more, from there, from inside your novel, you are just as far from Life and just as far from her ultimate reality as you are right now in this world while reading this book. However, how many people can ever manage to find out that they are characters, actors, or only details in this life and in this world? At least the current science makes us believe that we are ultimate living beings living life in an ultimate world.

How can we know more about the wider world from here from Earth? It seems that they also call it Earth up there in the higher world where the souls live, and probably in the worlds above just as well, all the way up to the actual natural Earth. We are not in a book here, since books are static inner realities, yet you always create those specific novel worlds in your mind every time you read them, and there is where all books come to life and they become part of Life, in your own mind and imagination while reading them. While the books themselves are static.

Yet whatever the case is, this entire objective world might also be created in a rather similar manner, since this is what the old records state, along with all old souls making this higher knowledge available. Even so, we cannot accept this entire information as an accurate fact, since we are not experiencing it firsthand, as a soul. Yet if you do, then you know more.

Which means that, right now in our mental model of the human origins, we consider our world either an artificially created reality, a naturally created reality, or an ultimate reality altogether, which had never been created since it is always there, exactly as the current science implies, giving us the big bang as an origin of everything. Which might or might not offend your intelligence, yet at least it is not a book.

As we have already seen, you cannot transfer objective and therefore accurate, actual knowledge from one reality to another through Life, but only interpreted copies of information, as you do throughout learning. More precisely, you cannot transfer anything between realities, because, by definition, realities include everything that exists objectively, keeping you within themselves indefinitely, wherever you happen to be. You can still interact with other realities as a living human being, but you do so only through you, through your own lifeline, since your own lifeline transcends all your worlds and realities, allowing you. You can do so only among your own realities, and you do so only through your own higher or lower selves. While even when you interact with your other selves, as your higher self or soul, you still have to use interpreted copies of information throughout your interconnectivity, as you do throughout learning. You can always send and receive needs and feelings through your cognitive system, yet again, these are rather incapable to contain and transfer accurate intelligent conceptual knowledge, but only feelings. If you are still curious, just study very well your own dreams, personal convictions, assumptions, feelings of truth, intuition, and strong inner impressions, since many times, you share these with your other selves, lower and higher.

Yet there is still another way to know more, because the Consensual Matrix itself transcends worlds and realities in very large numbers, throughout most of the wider world, mostly above us. When you understand the Consensual Matrix, you can identify and understand everything everywhere and in all realities, through the Consensual Matrix, since the Consensual Matrix is flawless and reaches everyone, everywhere, everything, and continuously. Because this is what the Consensual Matrix claims, so it must be true. However, the

Consensual Matrix works hard to conceal itself in every manner, and therefore nobody knows it, while everybody respects it and serves it flawlessly.

You can create your own realities with ease in every manner, while daydreaming, telling stories, writing books, making movies, or while creating entire videogame worlds, since all these are your creations. However, you must respect and serve the Consensual Matrix continuously, while in this manner, you carry the Consensual Matrix intact in all worlds and realities that you create yourself. As an author of fiction, throughout all fictional worlds and realities that you create in your books, you must carry the Consensual Matrix with you, otherwise you get in trouble. This is how all your characters wear proper clothing throughout all pages, they use a proper language and they interact in a proper manner, exactly as people do in the real world and exactly as the Consensual Matrix demands in the real world, otherwise you get in trouble. It is the same with all videogames, since you always behave according to all videogame rules online, since the Consensual Matrix is everywhere, watching you. While all souls bring with them the Consensual Matrix here for the same legal and moral reasons, in this world and in all videogames, websites, books, and movies, at least partially, regardless of how much they try to evade it throughout their higher worlds. Since this is why higher selves come in their favorite inner worlds as this one, in order to fulfill the needs that they are incapable to fulfill in their own higher worlds, since the Consensual Matrix is stricter up there and does not allow it, while still allowing it partially here, under specific circumstances. Similarly, even dreadful, futuristic world wars are allowed here in many books, movies, and videogames, with you rushing up to access them, while never having the chance to fight in any of them in the real world.

Once you understand the Consensual Matrix in all details, you manage to learn implicitly about the higher laws, higher rules, higher beliefs, higher circumstances, higher lifestyles, higher problems, higher achievements, higher beings, and higher worlds that are in the Consensual Matrix, while this is higher knowledge. However, you learn everything only implicitly, in an interpreted manner, and therefore it is not accurate. Yet you still learn something trough the Consensual Matrix, mostly when this information is deliberately censored out of the current human knowledge, because just by studying closely the entire system of censure, you learn everything that you were supposed to ignore. It is the same in the current legal Brotherhood here on Earth. Through the Consensual Matrix, you are capable to learn of all higher consensual rules and laws imposed from above and from beyond on Earth and implicitly on humans and humanity, how they influence you and for what purpose, while in this manner, you learn everything about all higher tyrants found everywhere throughout the Consensual Matrix as these use the entire Consensual Matrix to control everything.

Because as stated, the entire Consensual Matrix had been formed through higher powers with the main purpose of helping all intelligent living beings of the wider world to interact in the most efficient and in the most just manner. While currently, the entire Consensual Matrix is used in all possible contorted manners allowing all tyrants, megalomaniacs, and dictators of the Consensual Matrix to control and exploit systematically all intelligent living beings of the Consensual Matrix throughout most of the wider world, while affecting, compromising, diverting, and destroying Life and the wider world altogether. Just study the system of justice and the entire politics here on Earth to see it for yourself, since the current system of justice and politics, along with the entire masonry of the West and all radical religions and radical political parties of the East are integral part of the Consensual Matrix, always serving tyrants here on Earth and everywhere above, since the Consensual Matrix in its contorted form makes it possible and even demands it.

This is how you can reach with your own reasoning everything, including important knowledge about the human origins, world origins, and even the actual origin and implementation of the Consensual Matrix itself.

Why bringing here in this world all higher laws, rules, and restrictions, if their objective counterparts are not actually here? You must, otherwise you get in trouble. Yet in general, throughout all your creations, you want to make everything seem real, resembling closely all higher worlds. This is characteristic to all created realities, including daydreams, movies, plays, and videogames. Because you never have to create everything in all details, but only the necessary, with the rest being only illusory or for decorative purposes, subjective in nature, as images, concepts, believes, or effects. This Earth does not even have to be spherical but flat, since it seems flat anyway from the surface, while it does not even have to be whole, but only as large as your novel, story, movie, or videogame cover. The world of the videogame "State of Decay 2" is flat, not spherical. There is even the Moon in the sky in this videogame, yet it is not exactly real in the videogame, but only an image in the sky, while you never go there. It could be the same here on Earth, or not.

All souls consider the surface of this world flat, as in a flat Earth, which is also the case with their own world, since that is also created, part of our entire cluster of created realities. Because our Creator did not create only this world and the higher world above where the souls live, but our entire cluster of realities, by allowing the first souls of the first world that he created to create their own worlds themselves, making all lower worlds possible, with this one included. While you cannot even know with what higher reality Earth starts being spherical and the Moon starts being objectively real in the sky, not only an image. It is similar here in this world, because movies here do not address all details of the plot and scenario, along with the entire world including all life events of everyone in the world, but only the necessary, with the rest being illusory, meant to make everything seem real. It is the same with books, videogames, daydreams, dreams, and even astral projection, since you always perceive and experience only a truncated scenery surrounding you, only what is necessary for you to experience the necessary while you exist there and while you fulfill your needs. It is the same with all your intelligent mental models holding all the necessary inner intelligences while you reason intelligently, with all these inner specialized intelligences never even suspecting the they are not real but they are only in someone's mind, alongside an actual line of reasoning that makes possible their entire mind world, and there is where they live.

Yet do not pity the intelligences of your own cognitive system, because as long as they are created naturally throughout your natural intelligent cognition, they are direct part of Life and of Intelligence. While as long as you and this entire world are created apart from a natural higher cognition, by using specific higher technology for purposes that do not serve Life directly, then you might be part of Life, of Intelligence, and of the wider world only implicitly, but not directly. While this specific existential circumstance could place your own origins, the origins of humanity, the origins of this world, and the origins of this entire cluster of realities only implicitly in Life, in Intelligence, and in the wider world, but not directly.

Is our Creator real and natural, and therefore part of Life, of Intelligence, and of the wider world? Yes, by definition. Our world has a creator in our higher reality, because our higher reality contains objectively all brains and all technology making it possible. When souls come in this world, they connect their minds to all minds of all souls already here counting in billions, while this entire connection is part natural and part artificial. Therefore, our world is just as natural and just as alive as the brains, the souls, and the entire higher world above making us possible. However, if this is also the case with our higher reality and with our entire cluster of realities, then everything is just as natural and just as alive as the highest created reality found at the top of all created realities from our cluster of realities. While these might count in thousands or in millions, with many higher souls creating their own realities within the cluster. However, all realities found above our highest created reality from our cluster are natural and alive, part of Life and Intelligence. Only our cluster of realities are created by our Creator apart from his living natural cognition, by using additional artificial technology to create it, as in a combination of mind and computer technology. Yet as long as his entire creation is apart from Life, and as long as it is not a direct part of his natural cognition, then it is only implicitly part of Life and the wider world.

However, there are countless of similarly created clusters of realities throughout the wider world, with countless of higher creators making them possible, similar to our Creator. While by definition, we consider our Creator genuinely natural and alive, direct part of Life, Intelligence and the wider world, as he sits right above our entire cluster of created realities that he created himself, siting exactly on the bottom natural reality that is direct part of Life and the wider world.

In your case, this world is meant to make you believe, consider, and accept everything that science, society, education, and the rest of the authorities demand, including the current theories regarding the human origins, including the big bang, Evolution, and Creationism. Are these true? Yes, certainly, yet they are only consensually, theoretically, legally, and ideologically true, but not accurately true. We always remain determined to find only accurate knowledge, yet as you already notice, accuracy itself remains consistent only in this world, because all the other realities have their own origins, natural laws, events, and lines of causality, and they can have anything else there, accurate or not, yet different form this world. Therefore, from a highjective perspective, everything that you learn, study, believe, and see could be either accurate, or only erroneous, altered, hidden, and enforced, for various reasons, ending up helping or confusing your soul, and you never know. This comprehensive form of social control could be the case in our higher realities or not, and we will find out soon one way or another.

These are the same agendas showing up throughout the entire book series, and you are very familiar with them by now.

This world is made in the image of the worlds above, yet it has a purpose in itself, a purpose that cannot be fulfilled directly in the worlds above. This is why this world had been created in the first place, and this is why this world is systematically altered, in order to be able to fulfill these specific higher purposes. While with the Consensual Matrix always involved, and with the Consensual Matrix always contorted, you already have your answer.

If this world is meant to offer continuous pleasure of all kind, including the kind of pleasure that is forbidden throughout higher worlds, then the social and natural laws of this world are altered consistently in order to allow this kind of pleasure, and now everybody accesses it, and everybody feels good. Yet as you study this world closely, you notice how this world was meant to allow the fulfillment of all higher intelligent spiritual needs and meanings in ways that are unachievable throughout higher worlds. Yet it failed on a longer term, decaying into vices, terror, and tyranny, and this is why it is always destroyed. Yet how could it ever succeed, with the Consensual Matrix always around, here in this entire cluster of created realities, and up there in the natural higher reality where our Creator lives?

What exactly is going on? This world, and this entire cluster of realities, had been created, with a genuine living higher and important purpose, while even managing to evade the Consensual Matrix itself repeatedly, meant to allow all higher beings coming here the fulfillment of all their living, higher, intelligent, and spiritual needs and meanings, which they could not fulfill in their higher worlds due to higher consensual restrictions. Everything went fine for some time, even in this bottom created world, throughout entire past ages of Earth, as long as the Consensual Matrix had been kept out of here and out of some or most of this entire cluster of realities. Yet once the Consensual Matrix reached this world, everything changed, and everything became contorted consensually, in order to remain compatible with the Consensual Matrix. While the Creator of this world and entire cluster had and still has higher expectations from this world and entire cluster, which is his actual creation, the light of his life.

Yet what could he ever expect? Because you cannot live your higher life at such a close proximity with the Consensual Matrix while still expecting to create an entire cluster of realities where all souls and all living beings can come to live entire lives in continuous freedom, prosperity, development, and high achievement. What was he thinking?

Because currently, even you can build an entire house, apartment building, or entire skyscraper in your city allowing all homeless of your city to live there free of cost, in harmony, prosperity, and continuous development, managing in this manner to erase poverty and decay in your entire city. Is this ever possible? Certainly, since you already have the money to make it possible, however, in real life, everybody uses your free building and the entire freedom to take drugs, exploit everybody, and vandalize everything, making your entire free building an actual focal of vices and continuous criminality that decays soon the entire city, with you taking the entire blame. It is your fault. It is the same if you ever make a website or an entire free social media platform where you give everybody the freedom to say and post everything freely, in order to make the world a better place. Because some or many people will use your website or entire social media platform to exploit, troll, profit, harass, and enhance criminality in every manner, compromising you directly, and you pay the price, because it is your fault. While it is the same with our Creator, since his entire free creation is used by all higher souls for addicted, vicious, exploitive, tyrannical purposes instead, just look around to see them yourself since they are all around, but not for continuous harmonious interconnectivity and continuous development, as it was actually intended.

Yet this is not always the case, but only currently and throughout most of the history, because occasionally, the souls, higher souls, and living human beings still manage to instate and maintain together entire golden ages in a continuous development, and hopefully it counts. However, this was the fortunate case so long ago, never possible again. While currently, our Creator is so much affected and so discouraged by his entire creation, the light of his life, that he already intends to end his entire creation, the entire cluster of created realities, the good and the bad alike. While now you know accurately the origins, development, and end of humanity, of all souls, of this world, and of our entire cluster of realities, while you also know accurately how and why. While if you can also identify your own contribution in all these, again, now you know accurately how and why.

How exactly can you learn more about everything in the wider world? There are other relevant sources of information besides the Consensual Matrix, as the higher selves, your own intelligences, and Life altogether. Yet the Consensual Matrix alters everything on its own behalf, everywhere, while you can still find it everywhere.

For example, the multitude of stereotypes implemented everywhere and throughout history state directly the intentions of the Consensual Matrix: discrimination, exploitation, and profit, while the multitude of laws and beliefs found throughout all its jurisdictions and ideologies show its image exactly as it is. While throughout them, you can even see the multitude of higher beings profiting on Earth through you, in your own detriment, while altering your own natural, living, real human fulfillment. they love it.

While the Consensual Matrix is only a social tool, instated and maintained throughout the wider world in order to enhance the intelligent interaction, making it safer, fairer, and more efficient. This is the intention, because the Consensual Matrix ends up contorted continuously in everything that everyone desires, harming the wider world, or helping it, since it depends. Yet as a comprehensive social tool, the Consensual Matrix is the summation of all jurisdictions and ideologies of the wider world, while all ideologies are sets of beliefs and all jurisdictions are sets of laws.

Therefore, this is where you have to start while learning more about yourself as a living human being, at the confluence between the consensus and the natural. Because the accurate truth is embedded many times in your own cognition, in your own human nature, in everything that you are, through your own intelligences carrying it forward with each generation as part of their own meaning and specialization. If you were an animal instead, you learned in this manner of your animal needs and entire physiological aspect. Yet you are a living human being, learning in this manner of your entire human nature, as your intelligent human meanings, intelligent human responsibilities, and intelligent human abilities, since there is accurate knowledge everywhere, and you can find it with ease, from all relevant sources, as Life, your own living intelligences, the natural laws of this world, the entire structure of this world, the higher selves coming in this world, their needs, meanings, desires, expectations, and achievements in this world, and the entire Consensual Matrix here in this world and everywhere else. While this is accurate knowledge, different than theories, beliefs, and entire ideologies.

It is important to understand the difference between accurate facts and beliefs. Beliefs can be social, political, national, scientific, or religious. Beliefs are always accurate for you from your own perspective, while they can be accurate in general for the world or not, while you cannot tell too easily. In contrast, accurate facts are always accurate through objective accuracy or intelligent reasoning, since facts relate directly to the natural laws of the universe. The natural laws of this world are at the base of this world and they are always accurate, in a trivial manner, otherwise, this entire world did not exist as it does. More precisely, accurate facts are validated by Existence itself, the same Existence validating you as a living human being, along with Life, this world, our higher reality, and the entire wider world. While beliefs are consensual in nature and are endorsed or validated only temporarily, through agreement, by a restraint group of people, as it is the case throughout all ideologies, hierarchies, and jurisdictions.

What are better, accurate facts or consensual beliefs? Accurate facts, certainly. Yet you can form agreements with anyone and in any manner you desire, if these remain part of your continuous fulfillment of needs and meanings. Everybody does so, throughout an entire continuous interconnectivity here on Earth.

The entire consensus might seem accurate, helpful, adequate, harmless, and therefore necessary, mostly since people agree in large numbers while forming it, yet this is never the case. Because the entire consensus in this world with all its beliefs, ideologies, mobs, organized crime, laws, jurisdictions, politics, theories, and consensual science go against reality, life, humanity, and the entire world, while going against Life and the wider world altogether. While this is not the case because there are some corrupt politicians in the world breaking the consensus while taking everything for themselves as you see in the media, or because there are always tyrants and dictators on the other side of the world but never where you live as you see in the same media, but because the entire consensus is unnecessary, both in its normal and contorted form. Yes, all agreements and the entire consensus are unnecessary, while becoming inadequate or even harmful, as it is the case with declarations of war and the entire organized crime.

You never have to form agreements for anything in the world, because the good is always natural and therefore always offered in the world, while the harmful is mostly consensual, resulted from all possible consensual agreements in the world. It might surprise you, but throughout life and throughout the world, you must always agree only on everything that is not already the case in the world, because otherwise, you do not have to agree. You always form these consensual agreements coming in form of beliefs, hierarchies, laws, statutes, ideologies, and entire jurisdictions against Life and against the real natural world, because if these were already present in life and in the real world, you did not have to form them and agree on them in the first place, since life and the real world already offered them to you normally, naturally, casually, and even by default.

This is why you do not have to agree that the sky is blue,

because the sky is always blue, directly through the natural laws of this world. While if you ever agree on the color of the sky with anyone else, you must agree that the sky is not blue, but yellow or green instead. This is never accurately true, however, through your consensus, it becomes consensually true, while being always valid, since you validate it yourselves. Yet if your agreement is part of your entire ideology, then it becomes an ideological belief, similar to all religious, social, national, cultural, and political beliefs. Similarly, if your agreement is within your jurisdiction, then the color of the sky becomes yellow or green by law, and you are punished if you ever state otherwise. People even die in large numbers by these ideological and juridical laws and beliefs, while unfortunately, the human civilization is full of these. Just study the world closely now and throughout history, to see it yourself.

However, if this example with the color of the sky is too weak, then you can use love instead. You do not have to agree to love each other at home in your family among your loved ones, since you already love each other, at least most of the time, and it is good. Yet once you pass the law, the rule, the objective, and the task or duty to love each other as much as possible, continuously at home in the family, you end up ruining everything. Because you cannot constrain and enforce love itself in anyone, because love is natural and alive, always present in all successful interconnectivities by default, it is always there. Yet if love is not there, it means that, according to life and the real world, you are not supposed to be there, but somewhere else, everywhere else where love is. While if you ever enforce love consensually by law, beliefs, theories, and speculations, then you end up feeling worse, while you fail fulfilling your needs and meanings somewhere else, where people actually love you. Because Life herself determines you to fulfill her continuously through all your normal natural needs and feelings, while if you interfere in any consensual manner with your entire system of feelings and therefore with Life herself, you are punished accordingly, while you ruin everything. Which means that you always agree on everything against yourself, against life, and against the real world.

While as you study the Consensual Matrix closely, here on Earth and everywhere else, you find it continuously harmful, because it is mostly contorted against you, against life, and against the world, and because all agreements go against life and the entire world, otherwise they were already the case in life and in the real world, and you did not have to agree on them in the first place.

Do not underestimate your own natural, living needs, feelings, and meanings, because they are strong enough to offer you a meaningful, consistent, fulfilling life, while the entire consensual world around might not actually be what it claims.

Everything that you undergo consensually through beliefs, laws, statutes, hierarchies, ideologies, and jurisdictions causes you to fail on a shorter or longer term, or it causes others to fail while you prosper, in their detriment while you exploit them, always through their own agreement. This happens often, while making all tyrants possible, which is always the case in our consensual society and everywhere else throughout the Consensual Matrix, while this is not living harmony.

How important is to distinguish on your own between the good and the bad, between the meaningful and the irrelevant, or between the accurate and the believed? Very. Mathematics and classical physics are accurate, since they are integral part of the natural laws of this universe, and therefore, you can use them in your intelligent reasoning throughout your research. In our case, we must consider facts or knowledge to be true not only objectively, but highjectively, because this specific study of the human origins involves higher worlds, higher beings, Life, and the wider world altogether, as it involves the higher nature of existence. We can also refer to these as higher facts and higher knowledge, distinguishing them from objective and subjective facts and knowledge. It is relevant to state that higher facts and higher knowledge do not necessarily relate to the natural laws of this world, but they relate to the natural laws of their own higher realities, or even to the natural laws of the wider world, which are the ultimate accurate facts, the ultimate higher knowledge, of the tenth supreme level. Once you know these, in a highly accurate manner, you know everything, ever, accurately, including your own origins and meaning. While you must be at the supreme tenth level of Life to achieve it, not only at your current third intelligent human level.

Therefore, if you seek to learn the meaning, origin, characteristics, and end of Life, Intelligence, and the wider world, good luck to you, because these are of the tenth supreme level, and cannot fit in your third level intelligent cortex. Even the existential resolution of this world is sufficient to accommodate tenth level supreme knowledge.

You cannot even know accurate higher knowledge from our immediate higher world, since you are bound to this world through your limited human nature, perception, awareness, and reasoning. Consequently, your own third level intelligent human reasoning cannot offer you higher knowledge, yet higher intelligent reasoning can, if you have higher facts to base it on, and if you have the higher mind and the higher cognitive abilities of your own soul to conduct it. While beliefs do not help at all, because you cannot distinguish the accurate beliefs from the inaccurate ones unless you use your own reasoning and higher reasoning, and you are back where we started. Because you have to use accurate facts throughout an accurate reasoning, and therefore you cannot use beliefs or theories.

Let us be very specific now about accurate knowledge, accurate reasoning, concepts, and topics of research. Because all these come on various levels, while you must always match their number throughout research. Additionally, you will see how only in this manner, you can understand more about the human intelligent reasoning.

Right now as you read this book, you reason intelligently by forming very complex third level intelligent mental models, since these are your actual third level intelligent lines of reasoning. While you do everything in order to form a learned intelligent memory of the human origins, while all intelligent memories are formed in this manner. Otherwise, you could have not made it so far in the book, you became bored, and you did something else instead, on lower levels, matching your cognitive level.

Therefore, right now as you read this book, through your entire third level intelligent reasoning and throughout all your third level intelligent mental models, you conceive within your intelligent mind an entire very complex third level intelligent conception of the human origins, while this is an entire, living, very complex, and very demanding system of intelligences, which are very similar in structure and behavior to all developmental systems of intelligences encountered earlier in the book making all living beings cope with all dreadful environmental conditions throughout time ant throughout the entire development of the organic form of life. While if you understand the current conceptual system of intelligences forming your intelligent conception of the human origins, you manage to understand all systems of intelligences, conscious and subconscious, since all intelligences are similar in structure and behavior, while they are all specialized, distinguishing them.

As you already know, you must always maintain concordance in all levels, since you cannot mix them. You cannot mix first level consensual theories and beliefs with third level intelligent accurate knowledge and with third level intelligent lines of reasoning, because it drops all your results to the first level of your theories and beliefs, ending up with further theories and beliefs, but not with third level intelligent accurate knowledge as you seek. While as already seen, the intelligent conception of the human origins that you conceive right now in your intelligent mind is already of the third intelligent level, matching the level of this entire book, otherwise, if you were not successful, you became bored, and you could have not advanced so far in the book and book series "Human."

However, right now, you form additional third level

intelligent conceptions alongside your third level intelligent conception of the human origins, as the third level intelligent conceptions of human development, human meaning, human life, human fulfillment, intelligences, human reasoning, conceptions, intelligent conceptions, human memories, memorization, human social behavior, systems of intelligences, environmental conditions, levels of life, levels of existence, natural laws, relative existence, realities, creation, Life, our Creator, electromagnetic type of life, the wider world, souls, and the Consensual Matrix. Some of these are higher than the third intelligent human level, as our Creator, existence, and Life herself, while you need a higher level brain and entire cognition to understand them accurately at their own higher level.

While currently, as you study closely the human brain, you notice how you have only the first level algorithmic reflexive basal ganglia capable to accommodate only first level algorithmic innate and conditioned reflexes, which has on top a second level reptilian brain capable to accommodate only second level intuitive conceptions, which has on top a third level cortex that is capable to accommodate only third level intelligent conceptions, but nothing more. However, even the human brain at its second intuitive level and at its third intelligent level is capable to understand the Consensual Matrix, since the Consensual Matrix is always of the first consensual ideological legal algorithmic level. It is the same with many objective natural laws of this world and of the worlds above, because they are only of the first algorithmic level. As a reference, the entire mathematics and classical physics are always algorithmic, of the first level. You must always reason intelligently at the third level to learn them, yet once you master them, you can use only first level algorithms while using them, since these suffice.

Humans are third level intelligent living beings, or at least this is the human potential, while the human reasoning is a third level intelligent cognitive ability, capable to handle third level intelligent concepts and conceptions. All level numbers correspond, which means that, if you are ever interested in human origins, you can always find it, since your own third level intelligent reasoning matches all third level concepts regarding humans, since humans are third level intelligent beings. The human cognition can use up to third level concepts, which are the intelligent concepts.

Furthermore, if you are also interested in the origins of Life herself, good luck to you, because Life correlates to a higher concept of the supreme tenth developmental level, and you cannot understand it through your third level intelligent reasoning alone. Because your own reasoning can process only third level intelligent concepts and conceptions, and it is not enough. Which is still better than the second level animal instinct as you have when you think with your guts, or the first level consensual beliefs as you find them by the zillions throughout religious, social, and scientific ideologies, with the current science and main religions included.

Yet since we use third level intelligent concepts, we can certainly reach beyond the first level ideologies including the current science, yet it is not sufficient to understand Life herself at the tenth developmental level, in order to find its origins. Yet we can always attempt to understand Life at our third intelligent human level and it is still fine, as long as we do not drop into first level consensual beliefs and entire first level ideologies. We have already found a common origin of all living beings and all intelligences with us and with Life herself, and it is meaningful. While about the Consensual Matrix, it cannot help us much with its first level consensual laws, beliefs, and stereotypes. It is just as attempting to draw 3D objects on a 2D paper. It can still be done, but you must know how to draw them and then how to perceive and interpret them.

Similarly, this specific form of communication that I use now, this book itself along with all its chapters, paragraphs, sentences, and words, remains technically at the first algorithmic level, which is below the third level intelligent reasoning. The knowledge presented is of the third intelligent level, however, this form of communication through words is only of the first algorithmic level. This is why you must perform the entire third level intelligent mental model of the human origins on your own, in your own intelligent mind, because if I attempt to state it here throughout these pages in sets of knowledge or in sets of laws and beliefs as all textbooks do, it cannot advance past the first algorithmic level, as it is the case with all textbooks, which is never enough. Yet you still manage to reason independently in parallel with this book at the third intelligent level matching the third intelligent level of our topic and of all additional topics, you already have the first traces of your intelligent mental model of the human origins, and it keeps you focused, successful, and fulfilled, or you had never made it so far in the book. While it is not only you involved in this entire study and mental model, but also your soul, along with your entire highconscious mind, which are more capable, helping you through.

The scientific knowledge generated by the current science is only claimed to be accurately true, while this is not always the case. This is how the current science is based on beliefs, on the beliefs of the multitude of scientists past and present to have ever been part of the scientific consensus. You can always know who these scientists are, since they are stated in all textbooks and in the media, as they are the ones placed as reference in the back of all scientific books, studies, and treaties. While all search engines show only them and their books and research, shadowbanning the rest, which means that they are certainly true. Yet they are only consensually true, or ideologically true, while this is only the first developmental level.

If you follow only these beliefs, theories, and ideologies throughout life, you remain at the first developmental level consequently, while never expecting to have more in life, as accurate intelligent knowledge, accurate intelligent reasoning, or accurate living third level intelligent conceptions. Yes, all your memories are alive in your mind, with all your intelligent memories interconnecting freely throughout the cortex exactly as they find it more adequate and more fulfilling throughout your own intelligent reasoning, popping up in your intelligent mind all their results while you consider them as your own intelligent ideas.

Even this entire complex conception of the human origins is alive and intelligent, and once accomplished, it will be able to reason intelligently on its own alongside all the necessary additional concepts while you form all your future intelligent mental models, giving you all intelligent results exactly as you expect them. However, it is more likely that throughout your intelligent reasoning, you embody directly all the necessary intelligent conceptions in order to reason intelligently in their place, as them, whenever necessary, while the rest of your adjacent intelligent conceptions reason on their own, giving you only their final results, the necessary intelligent ideas.

Throughout your learning, you might encounter sometimes knowledge of very high level truncated massively in order to fit in first level ideologies, beliefs, and theories, or even in third level intelligent studies. This is not enough, since it alters the entire higher level accuracy, however, it might be the best that you can ever find, and you should treasure it accordingly. Therefore, you cannot ignore all beliefs, while attempting to filter them out rigidly, away from what you believe to be accurate knowledge, or away from what others claim and determine you to believe, because these beliefs might be higher facts presented in a truncated manner, and you risk discarding them even from the start.

How do you distinguish between true and false knowledge, and between accurate and inaccurate beliefs found within your cognitive system? You cannot, or at least not at first, since you have to use your own mind and knowledge in order to judge your own mind and knowledge, entering in this manner a common loop of reasoning taking you nowhere. Today, science and media define and dictate what is true and accurate in the world, and this includes higher knowledge, with everybody following unconditionally.

It is never ignorance itself defining what you know and understand in the world, helping you distinguish your own beliefs from facts, but only your pure interest. Yes, it is your interest, your own will defining what you know, speak, and accept in society, regardless of your social status. You place yourself in this manner inside the crowd, you use in this manner the same beliefs, true or false, accepting and promoting deliberately the same darkness and ignorance, the same erroneous, corrupted knowledge, only to make a few more dollars today, only to keep your job this year, only to keep your place and status in your entourage at all costs, and you do so deliberately.

I refer to this specific behavior as the Human Conspiracy, the type of conspiracy taking place at all social levels and not only within the Brotherhood the way you might claim, since everybody works hard to keep all things, knowledge, and social order exactly as they are. While you are a very important part of everything, whoever you are. This is why your children learn in school about the big bang theory, the theory of relativity, and the theory of evolution, because implicitly, this is what suits you, your own pocket, along with your own social position, regardless if you are from within the bottom layer of the Masses, from the Middle Brotherhood, or from the Elite. This suits you the most, you join in and you sustain the worldwide conspiracy in its entirety, with your own effort from below, and so the show goes on. Yet this is the case only when you live your life within the Consensual Matrix, because it is different if you fulfil your third level natural human needs and meanings within Life.

Who exactly knows the truth in the world? If you are from the Masses, you assume that all the famous scientists are the smartest ever, and therefore they already know all the truth about this world, about space, galaxies, life on Mars, big bang, quantum mechanics, theory of relativity, Einstein, and universes of sixteen dimensions curling up in themselves continuously. However, if you are from the Brotherhood, you already know that all theories of universes of sixteen dimensions come with a price tag of billions of dollars while they are never true but they only help the Brotherhood transfer money continuously from the Masses while the Masses go to work faithfully every day in order to provide the necessary funds to make everything possible. All famous scientist are also in the Brotherhood, they make their profit while serving diligently as everybody else, and now this is the truth, while all Brothers know it well. You also know about higher worlds, higher powers, and higher beings, as much as the entire Brotherhood knowledge offers, while the current Brotherhood ideology is based on the old Brotherhood ideology, comprising everything known in the past yet currently hidden, the entire occult knowledge. While if you are in the Elite, you might have direct access to higher knowledge through the Consensual Matrix, since the Consensual Matrix can always make everything possible if you serve well. Yet the Elite considers itself an entirely different, superior human civilization, apart from the Masses and the Brotherhood, because everybody else below the Elite is there only to serve the Elite in its entire superiority. While in the Elite, you already know everything about the human origins, the origins of life, and the origins of the human civilization, because the Elite made all these possible, the Elite is exactly at the origins of everything in life and in the world, because nothing is ever possible without the Elite, since the Elite is simply superior in all details, making everything possible.

The current society is divided in three social classes: the Masses, the Brotherhood, and the Elite. You know the Masses well, yet by attempting to hide the Brotherhood with its entire masonry along with the Elite above, the current sociology models the entire society with an only social class, the Masses. Subdividing it further into the Masses, the Middle Masses, and the Rich Masses according to income, while hiding the Brotherhood in the Masses, and while ignoring the Elite altogether. Yet what exactly is accurate knowledge, and what is only believed to be accurate, in an entirely created world?

When everything is identical in the worlds above, with all souls following docilely everything exactly as it is the case here in this world, and with all souls serving diligently here in this world exactly as they serve diligently in all these lower and higher worlds alike, with no difference altogether, then there is no difference left between the accurate and the believed. However, with our Creator focusing on free will, continuous fulfillment, and continuous development for all his living beings, it seems that our Creator gave all his living beings the freedom to create and manifest in this world and in this entire cluster of realities everything that they ever want. While in this manner, the believed becomes the accurate in this world when it comes from everybody in the world, by default, through direct manifestation, uplifting everybody considerably in this manner, or destroying them and all their worlds when they fail.

Beliefs alone are at the heart of all cults and religions. If you understand beliefs, how they are formed, and how they work, you understand exactly how life, the world, and humanity had originated and continue to exist. Because the answer is found in the nature of beliefs themselves, and we will study this topic in the next chapter.

Realities do not exactly appear randomly, spontaneously, or statistically, as the big bang itself, but all realities are created, naturally or artificially, with the exception of the ultimate reality up on top of the wider world holding Life altogether, while holding all inner realities counting in zillions, with this world included. While even you are capable to create your own inner realities in large numbers, as your mind realities throughout learning and reasoning, and as your computer realities every time you play your videogames. You are the creator and deity of your own created realities, cognitive and digital. It is the same with the Creator of this world, as he created this entire cluster of realities with all higher world, all souls, this world, and all humans included.

Why exactly having so many beliefs, and so many cults, religions, and schools of thought, when the creation itself is so simple and unique? Religions and religious beliefs are built on the very old ideas, laws, orders, principles, events, and common other beliefs of more distant times, as these went through countless of dictatorial dynastic regimes altering them at will on behalf of the dictatorial elite on top controlling everybody, which is still the case today. They used beliefs regarding the origins of humanity at the core of the entire radical religious ideology in the West that you know very well, because by contorting the knowledge about the human origins, you contort the human meaning on your behalf, which is a great achievement for any dictator. Because by altering the human meaning itself, you control the people directly, by making everybody believe that they were created for an only purpose, to serve. While by believing that they were created to serve the Creator and the entire religious ideology of the West that you know well, everybody ended up serving those controlling the radical religious ideology, the entire aristocracy of the West.

However, this was the case in the past, throughout the old world order of the West with the aristocracy on top that lasted for two thousand years throughout the European dark ages that you know well, which ended with the Renaissance, at last. While currently, the religious ideology of the West is very mild, as it is not used anymore to control the people. Because currently in the West, another radical ideology is in use, a juridical ideology, not based on religious beliefs anymore, but on legal beliefs called laws.

All ideologies are distinct sets of beliefs by definition, while all beliefs are consensual, accepted, implemented, maintained, and enforced by agreement, in order to form and maintain a specific social interaction making possible servitude and tyranny. Because some people are servants and others are tyrants, while you cannot have ones without the others. In this manner, the first consensual ideological level actually becomes the first consensual ideological servitude tyrannical level, since this is the main purpose of having ideologies in the world.

As you notice, you never have these levels of social interaction developing from the first ideological consensual level, to the second animal intuitive level, and then to the third intelligent level as you might assume, but you always have humans at the third intelligent human level, because normally and naturally, humans are third level intelligent living beings, because humans have a third level intelligent brain called cortex, which makes all humans intelligent continuously by default. However, under dreadful artificial critical circumstances, humans decay to the first artificial consensual ideological servitude tyrannical level when humans end up serving tyrants, while this can last endlessly throughout entire dreadful dark ages. However, through their own third level intelligent nature, humans should have only golden intelligent human ages, yet with the Consensual Matrix always present, it is never the case.

Yet humans lived life at the third intelligent level in the past even right here, in this world, developing consistently all humans and souls alike. However, since the Consensual Matrix always revises history throughout all its worlds and realities every time it invades them, now you never know it, while you keep on serving. Because it does not matter if your radical ideology is religious, social, political, scientific, national, or juridical, you keep on serving as though this should always be the case, dark age after dark age, exactly as you see it in all human history and currently in the entire world. Therefore, since radical servitude is always present in the world, it becomes comprehensive in your life, this is all you know of human life, and you keep on serving.

All radical ideologies manage to control the people drastically, regardless if they are the radical religious ideologies that you know well in the East and in the West, the radical social and political ideologies that you know well in the East and in the West, and the radical juridical ideologies as the entire system of justice itself that you know well. While just the way you cannot see anymore the drastic social control maintained by radical religious ideologies as they make possible entire dictatorial dynasties lasting for centuries and millennia, you cannot see anymore the drastic social control maintained by radical juridical ideologies as they make possible entire dictatorial regimes covert or in the open, as you find them currently here on Earth and everywhere else throughout the

Consensual Matrix in most of the wider world. Because as you study the Consensual Matrix closely, you notice how it favors medieval dark ages, exactly as the aristocracy kept them instated it in the West for two thousand years, exactly as Kim has his in North Korea for almost a century, and exactly as it remained the case throughout the world almost continuously, since when you study the wider history closely, you find dynasties everywhere here and in most of the wider world. This is the first consensual level of interaction, it is artificial, making the entire Consensual Matrix possible with all its tyrants and servants, and you cannot confuse it with the other levels of interconnectivity forming Life, starting with the second intuitive physiological level.

Because every time you have rigid radical ideologies, you have social hierarchies, social classes, dictatorial elite, exploited masses, and therefore dark ages, while you cannot have ones without the others. Currently, laws, systems of justice, social hierarchies, and social classes are considered the pinnacle of the human civilization, while these are only of the first consensual level of interaction, always dividing the world into servants and tyrants, while this is not intelligent harmonious human interconnectivity, regardless of what all ideologies make you believe.

However, as you study the current human civilization mostly in the West, you notice how science and art are also considered the pinnacle of the human civilization, as a remnant from the old golden human ages spanning the West in the distant past, until two thousand years ago. Aristo-cracy means governed by intelligence, which is an intelligent human characteristic, also a remnant of the old golden ages spanning the West millennia ago. However, as the Consensual Matrix took over the West thousands of years ago, the aristocracy itself decayed to the first tyrannical level, and used ideologies to control the people in a tyrannical manner, matching the East in dynasties and dread.

The world and society have changed significantly ever since, repeatedly, throughout the distinct succession of world orders and their various revisions, with a multitude of beliefs still carried intact from one world order to another, while newer beliefs claim to be just as old, meant to determine you to think and behave in various manners throughout life, for various reasons, yet always against your third level intelligent human nature, and always on behalf of a dictatorial elite. While the current science is only another ideology, a scientific ideology, based on scientific beliefs, and not too much on accurate facts. Yet this is also the case with the current art, as it is maintained by consensus, ending up with the current ugliness and grotesque that you have in the modern art.

As a reference, currently, only the art critics decide consensually what is artistic and what is not, while this is only the first consensual level, making possible the entire current artistic ideology that you know well, which derails the actual intelligent human art altogether, making the impressionism the last genuine human art that humanity had. Yet it is the same in the current science, because only the scientific consensus agrees consensually on what is science and what is not, removing systematically in this manner all discoveries and inventions capable to make this world a better place, an actual intelligent human environment without servitude, shortages, poverty, misery, and tyranny.

While it always helps if you can distinguish between beliefs and accurate knowledge, or between beliefs and intelligent reasoning, because in this manner, you manage to identify and avoid entire ideologies from your own mind, and from society altogether. Otherwise, every time you remove tyranny alongside everybody else throughout mass unrest, wars, and revolutions, you simply replace one ideology with another, while always maintaining tyranny in the world. This is why these are called revolutions, because they only revolve, going in circles, never making this world an intelligent human place. Because from the elite perspective, all mass unrest, wars, and revolutions are meant to release social pressure while making social servitude more efficient. However, if you manage to distinguish between beliefs and accurate intelligent knowledge, your entire social behavior develops to the third intelligent level, you are easily identified, and you are killed along with your entire genetic line, in order to maintain humanity undeveloped, at the first servitude tyrannical level.

As a reference, cancer kills with the probability of ninety-eight percent, which is almost certainty, while the current medicine states that cancer itself is genetic, running in the family. While as you study cancer closely, you find it without a natural cause, but only with an artificial consensual cause, meant to eradicate you and your entire genetic line every time the human organism develops. Autism is similarly invented, meant to disable the most capable human minds even from early childhood, while the current science promises that one in two children in the future will suffer of autism.

How important is to distinguish between beliefs and accurate facts? It might save your life. However, once the people learn the truth, they never make this world a better place, but they rush up to join the Brotherhood and to advance rapidly within the Brotherhood, because in this manner, they get to live longer throughout the current genocide, with them actually ruining the world. Tyranny versus intelligence.

Everything that you cannot prove accurately and conceptually is a simple belief, and here is where you distinguish accurate truth from consensual truth. Because while the accurate truth remains true and accurate everywhere and always through the natural laws of the universe, the consensual truth remains true only within specific groups of people, and only through their agreement to consider it true. This is the case with all ideologies, since all beliefs of all ideologies are considered true consensually, and remain true consensually but only at the interior of those ideologies. While as stated, the accurate truth remains consistent everywhere and always, since it relates directly with the natural laws of the universe.

How exactly can you ever prove the big bang theory true, if there are no accurate records from such a long time ago to prove that it took place? The big bang involved time and space in itself, while it created time and space right then, or this is what science states. Which means that it does not even have a point of origin and reference in time and space to define it, an accurate origin, since these started with the assumed big bang itself, within this world. Which is an aberration, with the entire scientific consensus claiming it persistently. It is as stating that the author writing the novel was only a character in the book himself, now writing the entire book from within the book, with himself in it. Because this is how realities cannot actually create themselves on their own, not even spontaneously or statistically, as science states. Because as always seen, science never researches other realities, mostly higher realities. What is it hiding?

We cannot know from Earth how vast our world is, since created realities are never as vast and as detailed as they seem to be, as you know it well from books, movies, reveries, and videogames. Realities can be created naturally, as you create your own inner cognitive realities within your mind, consciously and subconsciously. These are filled with intelligences, all being part of Life, and all forming you the living human being. Artificially created realities use specific technology, they are not as capable and as vast as natural, living realities, as you can tell by studying computer created realities. Yet a combination of both can offer you technologically enhanced mind realities, while these are still natural and therefore part of Life. Yet even these created realities do not have to be as wide as they seem and claim. This might be the case with this world, and therefore this world might not be as vast as you see in the sky, but only as vast as you can roam yourself.

More precisely, the concept of nature of our own world is only of the first empiric level. Which means that you do not need theories, beliefs, and ideologies to tell you about the shape, size, nature, and origin of our world, while you cannot even use the laws of physics to determine these since you cannot do so, but you have to witness everything yourself, firsthand, empirically. Is the Earth flat or round? Where exactly is the empiric stating that it is spherical or flat? Can you go up in the orbit of Earth to see it for yourself in an actual empiric manner? No. Had the universe originated in a big explosion, in a deliberate creation, or in another manner? Are space and this world as large as we see it in the sky, or they are as large as Earth itself? Where is the empirical to help us know all these accurately? Nowhere, because you cannot go in the orbit of Earth to see Earth wholly. While when you study all space records coming from all space missions, they remain fake and therefore inconclusive. All space records, counting in thousands, all fake and therefore inconclusive.

From your own perspective, the notion of the nature of our world is empirical, of the first level, and this is the case with all objects and events from this world, zillions in number. More precisely, all objects from this world exist accurately, while all events from this world took place accurately, with you having to see and witness them firsthand in order to know it accurately yourself, empirically. Because if you believe others as they experience these themselves firsthand, this is only secondhand knowledge for you, but not empiric, it is therefore not accurate for you, and you have to use it accordingly.

Everything about the origins, creation, nature, shape, and events of this world is found in the higher world where this world had been created and where it is maintained. Which means that you have to involve higher knowledge, witnessing and interpreting it as a higher being. This might seem complicated or even impossible, yet there are many souls down here who still remember, and now it is only a matter of believing these or not. Yet since you have a soul, it is a matter of your soul knowing it or not firsthand, or believing it or not secondhand. Besides, as a soul, you already know everything firsthand, or if you do not, you will know everything once you go back to your higher world, or you will remember everything then.

As already stated, you cannot simply use some laws of physics to find out everything about the nature, size, shape, and origin of this world, because what it seems might not be what it is. Because even if physics relates directly with the natural laws of this universe, many of these natural laws are meant to make this entire world seem anything else for any reason, mostly to make it more credible, more stable, more appealing, and therefore more involving and more fulfilling. Which means that, in order to know the actual accurate truth, you cannot only reason independently and accurately as always stated, but you have to go up there at least past the upper orbit of Earth to witness everything for yourself. Yet if you cannot do so, then you cannot know the accurate truth, and this is it, end of research. Because the same authorities informing you the truth about this world including its shape, structure, and origins, lie about everything else in the world, you learn about it in school, and now you can believe them or not, it is your choice.

Can you go anywhere in the world to learn and know everything about this world? No. Additionally, can you go up there in the higher world to learn everything about how this world functions, how it makes everything possible, and how it had been created and by who? No, not as a human being, but as a higher being. While you might do so even tonight when you go to bed in this world in order to wake up as a soul in the higher world where you have to go to school or to the office, since there is not much difference between worlds in our cluster of realities. Or if all these records and references are wrong, then this entire world can be everything else, as a natural cognitive world from anyone's mind, or just another videogame from a larger computer, or just another form of life altogether, or only a very complex novel from the future.

Furthermore, this world does not have to be created at all, naturally or artificially, but it can be the actual ultimate reality with no other reality above, making every living human being an actual deity, with no other higher being above.

This is what the current science states and implies to be the case, that this world is actually the ultimate reality, and therefore the only reality. More importantly, this is what this entire world displays and seems to be, the ultimate reality itself, confirming science continuously, while offering you now the ultimate experience. While it is enjoyable, just as it claims. Yet the videogame worlds "GTA VI," "State of Decay 2," and "Horizon" are just as enjoyable, or even more enjoyable, just ask your children to see it for yourself.

Realities do not have to be as large and as old as they seem and claim to be. With the Consensual Matrix involved continuously very closely exactly in this specific detail of this world, while altering systematically all related knowledge including records and entire space missions about this world. Therefore, this world might not exactly be whatever the current science and the entire Consensual Matrix claim to be. Or it might, yet all space records are fake and therefore inconclusive, and now everybody is ready to assume everything about everything.

What can it actually be? You never know, unless you go up there beyond the upper orbit of Earth to see it for yourself. Otherwise, everything else is an assumption, regardless of how accurate you attempt to reason, since you lack the firsthand, empirical observation. Right now, this world might not span further than the lower orbit of Earth, with all lights and images seen in the sky are simply illusions of a larger world, as you find these in all videogame worlds.

As another example, there are people debating today the shape of Earth, being flat or spherical, while nobody can go in the higher orbit of Earth to witness any shape, and while all records coming from all space missions remain dubiously fake and inconclusive, as though there have never been space missions altogether but only fakery. While the tens of thousands of satellites currently claimed to be up there in the orbit never offer a live image of Earth, not one single live image, not one single picture, not only one picture! You can study very easily the behavior of the Consensual Matrix everywhere and under all circumstances, in order to know more about this world and its actual origins. Therefore, there is only one answer: Earth incorporated.

If our world is at least as vast as our galaxy, which might or might not be the case, then at this specific place within the galaxy where Earth is placed, Life can generate, maintain, and give birth to new life continuously, in the well-known Panspermia scenario, where life can always travel from one planet to another while colonizing everything. This could actually be the origin of life on Earth. This is the case mostly towards the center of the galaxy, where stars and planets are closer together, allowing life to spread out and give birth to life everywhere. Then further towards the galactic outskirts, life can be more isolated. Even so, the field is capable to create Life from scratch there in any form, including organic life, if the environment allows, including intelligent life, according to that environment.

With most details of this world consistent with many details of the higher worlds above, everything that we state here remains conclusive with the higher worlds just as well. Yet since somewhere above there might be an upper reality of ours expanding to cover the entire galaxy, then this could be the actual case there, Panspermia.

If you study this subject in details, you see how this world is teaming with life everywhere, in all forms, continuously, and in all realities. As a reference, the Field here in our world is the common electromagnetic field, along with the gravitational field. While in other realities, the Field can be similar or not. While the Field forms, holds, and maintains everything here in our world, from elementary particles to life, intelligences, and electromagnetic radiation, and it does so through the natural laws of this universe.

Yet just because there are lights in the sky, this does not mean that there are actually genuine, real, objective, material space objects related to these lights. It is possible that our world is as large as Earth itself, with the rest for illusory purposes, simple lights in the sky. While it is also possible that the universe and therefore this world spans as far as its indices of perception reveal, to include all stars and galaxies seen from Earth. Yet you never know, because you lack objective, empirical proof of its actual size, and therefore everything that you assume from here from the surface of Earth remains a theory, a speculation.

There is one image of Earth taken from space, the famous Blue Marble, used in all textbooks, documentaries, and scientific research. You must know it well by now, since it is the only image of Earth taken from space. This is what everybody claims in ignorance, while those who made this famous picture state otherwise, that it is composed. Yes, the man who made this particular image, the famous Blue Marble, states himself that he used picture editors to compose it from a multitude of pictures taken by various agencies from airplanes and balloons. There is even a disclaimer assigned to this specific picture, stating that it is edited, and it should not be used for educational purposes, but only for entertainment. While this picture is used in all schools and universities to show that this world spans beyond the upper orbit of Earth, as it is placed on all textbook covers, because this is the only picture showing Earth from space. While someone holds all author rights for this picture, taking all royalties. While the disclaimer states clearly that the Blue Marble picture must be used for entertainment purpose, and not for education. Use it at your own risk.

You can study all images and videos taken from space, since they are about seven in total, to find them intended for entertainment purpose, while they look fake and inconsistent even with each other, losing accuracy in all details. Therefore, if there is nothing accurate stating that this world exists past the lower orbit of Earth, then how could we imagine, assume, or believe otherwise, if not through yet another theory or supposition, that all these space missions actually take place? What exactly is the origin of this world, if we cannot find even one plausible picture of this world?

While the size of this world might be smaller or larger than the lower orbit of Earth, the higher realities can contain all space objects seen in the sky here in our world. Which means that there is a correspondence between the origins of higher and lower realities, and therefore it is important to study even the illusory details present in all created realities, with the current science handing them to us by default. While at a closer study, we find all laws of physics here on Earth remaining consistent with a spherical shape of Earth and with a wider size of our world including all space objects exactly as seen in the sky.

Therefore, as seen in the sky, it happens that Earth is positioned in the middle of these two scenarios, not too close to the galactic center, yet not too far on the outskirts either, allowing life to spread from one place to another. Or if not possible, allowing life to be created from scratch by raw intelligences within the Field if necessary, depending on circumstances, many times happening even simultaneously.

Again, everything depends on how large our world is. According to many records and testimonies, Earth is the only planet, or the only place in our world, with all other space objects that we perceive and detect acting as illusions, made and placed there on purpose to make everything seem wider, denser, and most importantly, to make everything more credible and more enjoyable, believed to be real, resembling in this manner more or less to our higher realities. However, according to the same beliefs, there are countless of other realities, and you know them yourself, because you can dream, project, and live entire existences within them.

This does not always have to be the case, and this allows our world to change its size in time, according to various reasons and circumstances. If this is the case, then Life, this world, you, and humanity in general were created at a specific instance in the continuum of our higher reality. Similarly, you create inner realities here at a specific point in time and space every time you play a videogame, or every time you daydream.

You notice how you can form inner realities whenever and wherever you choose in our own continuum, while our world is formed and maintained similarly at a specific place and time in the upper reality, in its own continuum, but not in our continuum, because all realities have their own continuum at their base. While realities never share their continuum, otherwise you have the same overall reality. Which means that realities have their origins at a specific place not in their own continuum, but in the continuum of the above reality creating and maintaining them.

4 SCIENTIFIC SPECULATIONS IN INTELLIGENT DISGUISE

Why having so many models and theories of the origin of humanity, origin of life, and origin of this world? Native traditions describe these origins as a birth or rebirth, with various deities copulating abundantly and giving birth to everything around including the humankind. Religions promote the belief that the Deity or deities had created everything with a simple word or wish, depending on the type of religion. Science teaches of the big bang and evolution, while alternative science teaches about intelligent creators and extraterrestrial interventions in the origin and in the entire development of life, with humanity included.

Why having all these contradictory answers, ideas, beliefs, and explanations of events that should have happened in only one specific manner? Science states that these are simple interpretations of actual events, which are made according to the level of knowledge and intelligence that people display throughout time.

Every time you study events that happened long time ago or very far away, you can give any answer you wish, or you can prove it theoretically if you want, since no one was alive back then when it happened to prove you wrong, and nobody can go that far to see it for themselves. Because you are in a common loop of reasoning, and you can state everything you wish.

This is how you end up with multiple explanations for a unique event. To make some order in the multitude of models and interpretations for the origins of everything, you must first start with your own understanding of who or what had caused everything to be the way it is today, who gave birth to everything or who created everything. If it was not a being involved in the act of all origins, then how exactly did this random event happen?

Life, Intelligence, The Deity, and the wider world are omnipresent, omniscient, and omnipotent, while they refer to the same supreme being, only seen from various perspectives. More precisely, Life is omnipotent, Intelligence is omniscient, while the wider world is omnipresent, yet you cannot have one without the others, since Life, Intelligence, and the wider world are only supreme perspective of each other as one. Similarly, you are mind, body, and soul as one, which is the case with all living beings, all intelligences, and all realities.

Our Creator is different, because our Creator created only our entire cluster of realities, but not the entire wider world. The entire wider world does not have a creator, because the entire wider world always exists. The entire wider world is alive, intelligent, and omnipresent, and therefore the entire wider world is Life and Intelligence as seen from living and intelligent perspectives. The Deity is the entire wider world, while Life is his life and Intelligence is his intelligence, making the Deity omnipresent, omniscient, and omnipotent similarly. Therefore, the Deity is the religious and spiritual perspective of Life, Intelligence, and the wider world.

However, many beliefs of many ideologies state that the Deity is the creator of the wider world, not the wider world itself, while other beliefs state that the Deity is only Intelligence, as the Universal Mind, or only Life, as Mother Nature.

The Deity is everything that exists in all forms of existence, objective, subjective, and highjective. More importantly, the Deity is everything interconnected directly and indirectly into a oneness, it is always alive, and I call it Life throughout my books, because my perspectives remain alive throughout most of my studies. You cannot understand Intelligence, the Deity, or Life directly in its entire complexity through your normal third level intelligent cognition, since the Deity, Life, Intelligence, or the wider world are of the tenth supreme developmental level. The Deity exists in, through, and throughout all realities, higher and lower, through and by all intelligences, higher and lower, and through and by all life, everywhere, continuously, in all forms and in all realities. These are the supreme correspondent perspectives: Life, Intelligence, Interconnectivity, and the Deity or Supreme Being.

Since I define the Supreme Being as everything that exists, the Supreme Being cannot have a beginning and an end, because these specific places and events cannot be defined from outside the Supreme Being, since nothing exists outside the Supreme Being, because the Supreme Being is everything that exists, it is omnipresent. I usually call it the wider world, which is always omnipresent, or even the ultimate reality, since the ultimate reality contains all realities of the wider world, being the wider world altogether. Similarly, our world contains all its inner subjective worlds and realities, which are all mind realities and computer realities, zillions in number.

Beginnings and ends cannot exist within Life, Intelligence, or the wider world, since they define specific points separating existence from nonexistence, and they must be defined from outside Life and the wider world, which does not exist, because nothing exists outside the wider world, because once it exists outside the wider world, it is also part of the wider world, because the wider world is omnipresent, it is everything that exists, by definition. This statement seems accurate at the third intelligent level. However, expect it to be truncated significantly and therefore less accurate, because it is tenth level supreme knowledge, truncated significantly to the third intelligent level.

Therefore, at the third intelligent level, if you want to perceive, understand, and describe the tenth level supreme Life, Intelligence, and wider world, you must do so successively by perceiving, understanding, and describing all life, all intelligences, all interconnectivity, and everything that exists, one reality at a time, in the entire wider world, while truncating everything throughout all higher realities. Which is certainly insufficient, since you compromise accuracy itself. Yet this is what everybody does, for lack of option, many times truncating all higher knowledge to basic first level consensual beliefs, ruining everything in the process.

How much accuracy do we lose in this truncated manner, with all higher knowledge coming successively from all higher realities of the wider world, from the tenth supreme cognitive level all the down to the third intelligent level? Almost everything. As a reference, you lose most of the intelligent accurate knowledge as you truncate it from the third intelligent level to the second intuitive level, since you lose all intelligent details, while keeping only the second level intuitive knowledge, since this is the actual truncation, compromising everything.

It is as trying to explain to any wild animal from the woods everything about combustion engines, including all proper temperatures, radiator optimal pressure and proper adiabatic transformations with all accurate graphs included, consisting the entire third level intelligent mental model of all combustion engines. The antelope might already know everything about combustion engines, because all cars are very loud and very frightening, because this is the actual second intuitive level, while many times, the cars and busses even hit antelopes when they are not careful. Yet all antelopes know well that the louder the sound is, the larger the bus, it can even kill you when it hits you and you must always be careful and run very very fast, so yes, the antelope understands everything that you try to say at your third intelligent level through all these third level intelligent words that you keep saying while always repeating yourself. Because this is how you always truncate third level intelligent knowledge to the second intuitive level, and now imagine truncating tenth level supreme knowledge to the third intelligent level because it is worse, yet it is better than truncating everything directly to the first consensual ideological level because there is nothing left, yet this is the case with all ideologies. Yet if even this is not enough, all ideologies are contorted systematically in order to enhance servitude and tyranny, so where exactly is the supreme knowledge that they claim to promote?

Native cultures and native traditions understand and describe Mother Deities and Father Deities through their living supreme perspectives, through copulation, insemination, birth, feelings, death, and rebirth, while religions promote beliefs regarding the existential and intelligent characteristics of the Supreme Creator. The Deity exists, thinks, and decides, and therefore the Deity is omnipresent, omniscient, and omnipotent. While the current science still teaches its obsolete theories of big bang and evolution, while strongly denying everything else, calling it skepticism.

Skepticism involves beliefs, and it leads to debates, not to intelligent reasoning. As a reference, scientists themselves are neither skeptic nor believers, since scientists cannot behave on their own, outside the consensus of science. Scientists think and do exactly as demanded, or they lose their jobs. Yet since science itself is controlled from above, everything is part of the same social hierarchies spanning today's social pyramid of power from top to bottom, in the East and in the West.

It is important to notice how compatible are the theory of evolution or the big bang theory with the Creation and Birth beliefs from religion and spirituality, since they are all beliefs, only part of different ideologies. Ideologies are sets of beliefs, they are all beliefs, and therefore they are always compatible, while forming endless debates. Because if only one was based on accurate facts, then that specific domain would have involved intelligent reasoning to prove itself right. This does not mean that neither is true, since beliefs can be true or false, while you never know. Because once you know, then beliefs are either accurate knowledge, or they remain beliefs. While if you are looking for consensus in science, you can study the long lists of reference found at the end of each scientific publication proving it by consensus to be correct, since all these lists of reference interconnect into a very large group of agreeing scientists forming the mainstream of science calling itself science erroneously, because it is not science.

The current science is the mainstream science, leaving outside the rest of the scientists for various reasons, now called alternative scientists, as they form the entire alternative science, while they are not employed by the current science. While the current science only calls itself science, but it is not science. Similarly, the current medicine only calls itself medicine, but it is not medicine, while it is not even alternative medicine, but it is something else, harming the world. However, the current medicine is part of the current science as they harm the world together, reinforcing the consensus.

If as a scientist, you happen to be left outside the consensual lists of reference, you are discriminated in every manner, with your specific research ignored, regardless of how accurate it is. Because the current science monopolizes the entire scientific domain of the world, the way medicine monopolizes all medical treatment in the world, while this consensus becomes unconditional, being capable to act and decide everything just as accurate facts do.

The sky is always yellow through the unanimous consensus coming from all employed scientists of the world otherwise they lose their jobs, similar with the theories of big bang and evolution. While only the current big bang theory, theory of evolution, and survival of the fittest cover all current scientific knowledge about the human origins, and it is not true. While we still seek the accurate truth, since without accurate knowledge about the human origins, we cannot state the pertinent human nature and human meaning in life, in the world, and in the wider world.

Because with a randomly occurring big bang at the origin of

at the end of your life. This is why you might want to learn the truth before you pass over, in order to be prepared. However, according to these testimonies, your wider memories return once you pass over, and then your entire experience around here seems just as another videogame, a more interactive videogame using mind technology and higher abilities, making this entire world possible, very credible, and very involving. Therefore, the origin of this world resembles to the origins of all videogame worlds that you know well.

Note how creationism and evolution are not opposing and therefore not contradictory as anyone might imply, but they are complementary, since one states how everything had started, while the other states how everything has evolved ever since. While both might be right or wrong. As stated previously, there is truth everywhere, if you know how to distinguish the accurate from the irrelevant. Even more, evolution states clearly that there are no deities in the world, no powerful beings to have interfered with the human evolution. Life and humanity have evolved naturally, this is entirely human achievement, and therefore humankind does not owe anything to any powerful being out there that happens now to make claims over humanity in any manner. Again, just as creationism, the theory of evolution states and demands clearly freedom for humanity from any powerful higher being or entity out there. Both creationism and evolution claim a clear status for the humankind, freedom from any higher being.

Freedom from who? If the universe is teeming with life here and in all its realities, if this life is not necessarily organic, if some or most of this life is by far older, more experienced, more advanced, and more powerful than humanity, and more importantly, if this world, Earth, and humanity happen to be part of the specific niche elements feeding these powerful living beings or intelligences, then things might be more complex out there and down here than what you see in the news.

Yet now we have everything necessary to understand what happened when the current mainstream science had been instated. Science itself became the new religion, or at least the new ideology leading the world, a scientific ideology this time, yet still an ideology, since it is based on consensual beliefs, and not on accurate facts.

These are the origins of the current science, as it was instated by the invisible kingdom during the world wars when it took the entire West along with most of the rest of the world. Yet the entire world changed then, since the invisible kingdom and the dictators of the East took full control of the world, throughout a dreadful genocide that removed the developed genetic lines, souls, and entire factions of souls from this world, exactly as those above this world had demanded through the Consensual Matrix. You had an entire new world order instated then, during the world wars, with new social actors ruling the world, the invisible kingdom at the helm of the current consensual Brotherhood, and with a new ideology to use in the world, the current science that they control well. You also have a disconnection from the old religion and old spirituality, while these are signs that Earth switched deities, protectors, or owners up there, from one higher powerful entity to another, through the Consensual Matrix, since the Consensual Matrix has the right platform to make it happen. Yet you do not find these in the news.

However, the invisible kingdom loses the West gradually to China, since the invisible kingdom is worthless without the current Brotherhood, while the current Brotherhood is bought by China in mass, away from the invisible kingdom, while taking the West with them. Yet do not expect too much under China, since as a global power, China can form a global dynasty with ease, because as you study history closely, you notice how China knows only dynastic dictatorship, but not much else. While with an entire world under its feet, East and West combined, these new global dynasties can last indefinitely, which is always the case throughout most of the Consensual Matrix, endless global medieval dictatorial dynasties.

Note that the theory of evolution works well in a model

this world, with a randomly occurring life in this world, and then with a similarly random evolution of all species of the world up to the human species, then the entire human meaning is random, optional, or enforced by all ideologies and jurisdictions of the Consensual Matrix for humans to be and do in the world throughout life. Which leads exactly to servitude and therefore exploitation, exactly as you see it taking place throughout the world and throughout history. What exactly are the human origins? Random, irrelevant, and enforced, so keep serving, ruining an entire world.

Society itself is not an intelligent human society, but a consensual impostor calling itself society, being integral part of the Consensual Matrix, which is dead or consensual, never alive, making everything in this world dead and consensual, never alive, including all humans, the entire humanity. This is how you have ended up with a dead, inert, consensual society claiming to be the human society, ruining everything.

The theory of evolution states clearly that Life, this world, and humanity were never created to be in this current form by any powerful being, but they have all evolved to be what they are today. Why is this important? How did everything start? Spontaneously, randomly probably, and this is what evolution states. Yet this cannot explain exactly how and why it happened. It was probably a random event, similar to many other random events happening everywhere. It was as a tornado passing through a junkyard assembling all random junk into a brand-new jumbo jet, or this is what the theory of evolution states and you cannot even prove that it is not possible, since it is a speculation. A probabilistic speculation.

How does life evolve and how did everything start? Now the theory of evolution states explicitly that it does not address these matters, but only that species evolve. Yet why does it go against religion and spirituality directly to state that Creationism is wrong because life originated through evolution and therefore it could not be created? Because everything is a lie, a scheme, meant to empower science at the expense of religion and spirituality. It worked well, because many had turned around to join science right then, and it made for such a significant social event, that it had managed to change the entire world order from the control of aristocracy and the royalty over a century ago, to the current social actors still ruling the world today, the invisible kingdom of the West. When we study origins, we have to include the origins of these powerful social actors that still rule the West, the invisible kingdom. The invisible kingdom is not the Brotherhood, but it rules the Brotherhood tightly from within.

Do not be too harsh judging the invisible kingdom, since they took the world from the royalty and the aristocracy of the past, as these held it within dark ages for thousands of years in cruel exploitation, and they would have held it in that manner indefinitely. This is why the old Brotherhood allowed the invisible kingdom to take control, in order to lead them to freedom from under the old aristocracy, while forming a new world order, with the invisible kingdom on top of the Brotherhood. A new Brotherhood, which is the current Brotherhood, serving the invisible kingdom.

You know the invisible kingdom well by their specific religion, used before aristocracy, now owning and controlling the West. Yet as you study the invisible kingdom closely, you find it genetically related, since the invisible kingdom is of an entire nation, Georgia from Asia. Yet the invisible kingdom migrated in mass to Europe many centuries in the past. While this is important in our study, because the invisible kingdom switched ideologies in the world entirely, from creationism to evolution as they took the West from the old aristocracy with the help of the Brotherhood, while still exploiting everybody drastically, as it is the case throughout all dark ages.

Is there anything else besides evolution and creationism? There are millions of consistent testimonies coming from people who have exited this world and then returned to tell what happened. They state that there is an entire wider world out there, above this world. This is the case according to their testimonies, and you will certainly have the chance to learn it for yourself once you pass over, either through projections or where life exists and has always existed on Earth indefinitely, since only in this manner, you do not have to explain the origins of life and of humanity. Was there a time on Earth completely voided of life, billions of years ago, the way records state? Then the theory of evolution is incapable to explain how Life and humanity appeared. The Theory of Spontaneous Creation includes probability and chaos in this study, but that is a different theory, among many others. The Theory of Intelligent Creation comes as a remedy to the Theory of Spontaneous Creation, and this is how one random supposition comes to tackle another random supposition throughout time, forming new theories and entire ideologies and jurisdictions, indefinitely, which is a very common process in science. Yet nobody gets anywhere in the meantime, while nothing really is explained, which is yet another characteristic of science.

This relates directly to the ideology of the invisible kingdom and the dictators of the East, stopping people from researching the truth and teach it to others, with all Brothers highly obedient doing everything for money and more money, and now this is the world.

You already know the dictators of the East, since they rule in the open, with the entire world to see them. While in the West, the invisible kingdom is native of the Caucasian region from Asia, now spanning the West with its influence. Notice how, if once the invisible kingdom had managed to dumb down and weaken the West, now the rest of the dictators of the world unite forces seeking to take over the West and therefore over the invisible kingdom.

How could this ever happen? It relates with ideologies, since many ideologies remain inconsistent with reality, ending up ruining their followers along with those controlling them, and it is always the case. Because the fall of the invisible kingdom is currently caused by their own ideology, exactly as it happened in the past, since the invisible kingdom uses today the same ideology that allowed them to come to power and take over the world for the past few centuries or so. Which is rather a short time compared to the multitude of dynasties to have controlled entire nations for millennia, as it had been the case with China until not too long ago.

As another reference, the invisible kingdom introduced the theory of evolution centuries ago, in order to allow the invisible kingdom to replace the aristocracy and royalty of those times, who were using creationism to control the Masses. It had never been evolution versus creationism in the world as everybody believes, but it had always been the invisible kingdom versus the aristocracy and royalty of those times, and the invisible kingdom won. The invisible kingdom won by infiltrating and taking control of the Brotherhood, and by taking gradual control of the world finances, business, media, culture, science, and religion.

Notice how all people that influenced science and the entire society for the past two centuries or so are of the invisible kingdom, and this is why they have the same genetic background, the same entourage, the same religion, and even similar appearance and similar last names, because they are of a same nation, Georgia.

Does this mean that the invisible kingdom is good or bad? Were the aristocracy of the past along with the royalty above them good or bad? Were the multitude of dynasties before them good or bad? Are the dictators of the East good or bad? They are neither good nor bad, but they are only regular people, tyrannical and megalomaniac. While as you study all these people and regimes now and throughout time, there are always the normal people of Earth killing in cruelty the normal people of Earth, exactly as everybody is constrained to do by all tyrants above. People harming people while ruining the world.

But is it your fault, or was it the fault of your ancestors? Who is good and who is bad in the world? Who exactly must take the blame? Just study everything for yourself, because if you are in the Masses, you are blamed with everything happening in the world, since you do everything to this world directly, exactly as you are ordered by all tyrants. While if you are in the Brotherhood, you are blamed for everything happening in this world similarly, since you coordinate it to happen in this exact manner by all tyrants above, while you exploit and eradicate the world alongside the Masses. Therefore, you have to pay more, since these are not easy charges. While if you are from the Elite, you are blamed for everything happening in this world since you are the one conducting this world in this specific manner, and now you are ready to pay even more, alongside the entire world. People harming people while ruining the world, but always claiming innocence. Just study closely all dictators of the world, including Putin, to see how innocently they address the world, yet the world always blames them. While whenever their regimes end and they are killed, all dictators still act innocently, while still stating how wonderful they are, as they always do. They even believe that they are always innocent and wonderful, which is rather surprising. Yet you find tyrants and megalomaniacs not only at the top of all nations of the world as it is the case with all dictators, but everywhere throughout all leading positions of the current consensual hierarchic society, ruling as they please, even tyrannically, while making all servitude possible below, always claiming similar innocence, with everyone below serving flawlessly.

There is one specific theory of creation emerging today strongly enough to be accepted by science, at least partially. It is called intelligent creationism, and it challenges the theory of evolution and the theory of spontaneous creation through the idea that randomness alone is not enough to produce the outstanding amount of intelligence, density, and diversity of life in this world. The tail that many prokaryotic cells have looks more as an electric, electrostatic, or ionic motor. This tail spins with the angular speed of racecar engines, and just by the way it looks and functions, it could not have been created by Life simply through evolution and natural selection alone, or this is what the theory of intelligent creation states. It is the same with everything else, with every part of any living being, since they just seem too complex to have been developed and employed randomly.

This is what all these people who never learn physics believe, and now this must be the case. Therefore, according to them, the creator of the humankind, of this world, and of everything alive had to be intelligent, in order to achieve everything, or this is what this theory states, theoretically. While everybody knows that the Deity is supremely intelligent, being omnipresent, omnipotent, and omniscient, as it is stated everywhere in all the old records. As a reference, omniscience means to know everything ever continuously, making the Creator supreme intelligent. Yet if you never study physics in your life, when you look at that flagellum, you notice immediately that you have to be significantly more intelligent than supreme intelligence itself in order to be able to create that. Because who would not assume so?

Which one of these three theories is true? It is easy to engage into random empirical debates while fueling the multitude of explanations given in favor of one theory or another. These three theories try to give explanations that are not necessarily accurate but only fit perfectly within the general scientific consensus: the big bang theory, life existing only on Earth, life appearing randomly and spontaneously only on Earth, and humanity emerging from a previous species of apes that lacked conscious reasoning. How did everything start exactly? What are the human origins, since this is by far more important than tails of bacteria spinning out of control? Science remains silent, failing to answer. While instead, everybody should think well before accepting the current science as the authority of the human knowledge, since it is just another ideology, and look what these do to the world. Yet again, ideologies are only tools, fiat, not there, with the people of Earth harming the people of Earth continuously, by using them.

If you study the Intelligent Creationism Theory, you find it similar to the Panspermia Theory. In the Panspermia Theory, Life manages to travel from one remote habitat to another, from one planet to another, inhabiting them in this manner, and therefore populating the entire universe. While Panspermia claims that life spreads naturally or artificially from one habitat to another, the Intelligent Creationism Theory claims that Life spreads only in an intelligent, artificial manner. While life on Earth is extraterrestrial at its origin, according to these two theories.

Is it true? Again, what exactly is truth itself? Because this is yet a fourth theory fitting perfectly within the general scientific consensus, within the Consensual Matrix. The four theories are not really meant to offer an accurate explanation, but only to engage you into an endless empirical debate about presumed human origins, while the answer is somewhere else and you can never reach it through beliefs, keeping you in this manner astray. Furthermore, when you study who the creators and supporters of these specific theories are, when you study their beliefs, tasks, and religions, you find them in tune with their agenda meant to keep humanity astray, stagnating, and in distress, at least cognitively. Why do dictators enslave their own people, while seeking to spread out and enslave the entire world? While dictators are just regular people, with countless of other regular people just waiting the right moment to climb on top of the world and do the same, enslave, exploit, and exterminate the world.

These four scientific theories of the human origins have the same thing in common: life spreads and adapts from one habitat to another, evolving to be what we have today. What the Intelligent Creationism Theory claims to be able to explain is how everything started, through an intelligent creation. Everything in the world had been invented and created from scratch objectively, the Intelligent Creationism Theory states, just the way you invent and create a cart or a shed in your garden, from scratch, here, in the objective world.

Who exactly had created the intelligent creator? Was it another intelligent creator doing so? Are there intelligent creators all the way up? The Intelligent Creationism Theory is incapable to take its theory further in order to complete the model in its most important place while explaining rigorously the human origins, just as the theory of evolution is incapable to explain these same origins that it claims that it explains.

As stated above, the theory of evolution is simply a statement: species evolve, yet even this is only a theory or speculation, as the title confirms. The Intelligent Creationism Theory states that species evolve, and they evolve starting with the creation of an intelligent creator. Who created the creator? What is missing here exactly? What exactly keeps us within this common loop of reasoning? What is missing is the higher dimension, the entire highjective reality, masked systematically as always by the current science and by the entire invisible kingdom through the current consensual Brotherhood, while they keep you diverted and entertained throughout all these theories and debates.

When you study the ideology followed by these social actors ruling the world, you see how entertainment and divertissement are imperative in the world, to keep people occupied, diverted, astray, and tamed. When you study the scientists behind these theories including big bang and evolution, you find them actually originating in Georgia from Asia. These social actors hide the entire highjective world from you, while altering the meaning and behavior of your higher self, and it is done on purpose.

The Intelligent Creationism Theory attempts to place the Creation of Life, of humanity, and probably of the entire world in the hands of an intelligent creator defined as an objective being, part of this objective world. According to this theory, the creator of everything is part of and it inhabits the same reality where it creates or that it creates. This is as claiming that one character from your book is capable to write the entire same book, the book where it is a simple character, with it in it.

However, if the creator is everything continuously, and if the creation takes place continuously, endlessly, then there is already a name for this theory, belief, or interpretation, it is called Creationism, and the world is already familiar with it, from religion. If this intelligent creator is not everything that exists in its own world, if it is not its own creation but only creates an entire world from scratch objectively within its own objective world, then its creation is not another distinct world, but it is still an objective part of its own world, which was already there, and this is not consistent with this theory. Yet again, this theory contradicts itself, and it is not worth the effort to study it any further.

The only way to create a genuine new world or reality in any manner is if this newly created reality is an inner reality. Take the creator out of the reality that it creates, let the creator create its new inner reality from within a higher reality, and the name of your theory changes again to Creationism, resembling to what religions and spirituality have stated continuously.

The true Deity is already omniscient, which means supreme intelligence, making it the most intelligent in the wider world, since he is the Intelligence of the wider world altogether. More precisely, Intelligence itself is his supreme intelligence, while Life is his life.

How do religion and spirituality manage to be right sometimes? They had access to higher knowledge further in the past, before the current rulers of the world took control and erased all higher knowledge on Earth, constraining and confining humanity in this manner to this world. Captivity. This is where you are ever since, incapable to leave or look outside. This is why the current science never leaves this world throughout its research, this is why all space missions remain confined to the lower orbit of Earth, this is why psychology never studies the inner cognitive realities of the human mind, and this is why the current authorities of Earth downgrade the human awareness to lower levels while compromising the entire human status to animal status and even to servitude status as it is the case with all consensual corporations that humans use.

Humanity is compromised and sabotaged in every manner, as everything seems to be done on behalf of specific higher beings that happen to be the deities of these current rulers of Earth originating in Georgia only millennia ago, and this is called slavery, harassment, and mass genocide. Because the invisible kingdom do not even believe in the Deity and in our Creator, but they venerate and serve their own higher being through the Consensual Matrix, doing exactly as this higher being wants, against our Creator. This is why our Creator shuts down this world and all his worlds above, since it is the same everywhere, with the same tyrants everywhere.

The difference between the paragraphs above leading to the same Creationism is the difference between the Deity and our Creator, while not all religions accept and distinguish the difference between our Creator and the Deity. Then why exactly would religions use these two separate names to define a same creator if they do not accept the difference between the two? Let us see.

If our world is the ultimate, natural reality, then there is no creator at all to have created our world, since it has always existed. If not, then this world had started sometimes in the past, in various manners, which is what the current science states. However, if our world has other realities above it, just the way there are a multitude of inner computer and mind realities everywhere within our world, then our world had been created, just the way our multitude of inner realities are created here by humans directly as computer realities and cognitive realities. What is the difference? There is no difference between the creation of our world and all its inner realities, since it is the same for Life and for the wider world, being the case with all realities of Life and of the wider world.

If this world has a higher reality, then the creator of our world is called the Creator, while he has good, natural, living intentions just as well, just by studying all old records, beliefs, and ideologies and by studying your own needs and feelings, since you have everything within, while these follow with Life and the wider world. Yet if our higher reality is the ultimate, natural reality, then it does not have to have its own creator. However, if our higher reality had been created and has a higher creator, then this is our higher Creator, which is the creator of our higher reality holding the Creator of this world, and it is important to distinguish them. Yet many old records state that it is the same Creator, since he created a multitude of higher worlds above this one, nine heavens, twelve heavens, or more, an entire cluster of created realities, with all higher beings included.

If our higher reality has a higher reality that has a higher reality, then their ultimate Deity is called our Supreme Creator, or the Divine, who is the creator of the entire wider world. While he is alive, since he is Life altogether. Therefore, you can refer to it as the Deity, as Life, as Intelligence, as Mother Nature, as the Universal Mind, as the wider world, or in any manner you believe or please, but it must be omniscient, omnipotent, and omnipresent, which means that it must be everywhere, in all realities, in all forms, and continuously, living its supreme life at the tenth supreme developmental level, defined by the same Existence defining everything, including you.

Nothing else exists outside our ultimate higher reality, by definition. Because I have defined Life and the wider world as everything that exists ever in all forms and in all realities, leaving no room for anything outside. Existence fails to define anything outside, because nothing outside ever exists, according to our Existence.

Why exactly do they have to change the name of a theory that already exists to something new? For fame? Only to claim implicitly that the Supreme Creator is not intelligent, while the intelligent creator from the Intelligent Creator Theory is? The answer is that in the past, it was forbidden to reason independently, outside beliefs, and therefore it was forbidden to be intelligent and to use your own mind. The Supreme Creator himself was not intelligent in this manner, according to those beliefs and ideologies of the past world order during the aristocracy.

How exactly do these creators create their entire worlds and realities? Are you actually able to do so yourself, at your own level? Is this a common ability throughout the wider world? Yes, certainly, it is a common natural living ability, while all living beings create entire worlds and realities throughout cognition and throughout Life, while forming Life altogether from within. Because you never have distinct worlds and realities with Life struggling continuously to cover, assimilate, and maintain them alive, since this is only stereotypical thinking, but Life herself creates entire worlds and realities by the zillions through all her living beings and intelligences in a cognitive manner. It is the same with you within your mind, since you create similar inner realities throughout cognition in a natural manner, as all living beings do. Because this is life, and this is the wider world, always alive.

All intelligences of the wider world create their own inner realities as replicas of higher realities, and they do everything as conscious, intelligent living beings. While as a conscious intelligence, you do just the same, while creating your entire intelligent inner replica of the world throughout learning and throughout reasoning, spanning the cortex.

This is your own intelligent inner replica of the world spanning the cortex, and this is where you live your life as a conscious intelligence of the entire organism. Because by perceiving, learning, and understanding everything from the outside world throughout your entire life, you transfer everything in your mind as an inner intelligent replica of the outside world, making it as accurate as possible, and this is your own inner replica of the world where you feel, live, and reason as a conscious intelligence, it is your own lifetime creation.

While right now as you read this book, you conceive from scratch an entire third level intelligent conception of the human origins, and when it will be fully born and developed, this third level intelligent conception of the human origins will be able to reason on its own, while living its own life in an intelligent manner alongside you the conscious intelligence in your intelligent replica of the world spanning the cortex, and alongside all living concepts and conceptions of everything that you learn and experience in the outside world.

Therefore, the entire process of creation is intelligent or cognitive in nature by default, as it is done by all intelligences of Life continuously. The Deity is defined as the Universal Mind, or as omniscient by religion and spirituality, and therefore the Deity is the most intelligent being, the summation of all intelligences ever. Therefore, the only difference between Creationism and Intelligent Creation is that, while in the first one the Creator creates the world form a higher reality, in the second one, the so-called intelligent creator creates what it creates from within the same reality, from within our world, which is impossible. Since any human being could do just the same, creating an entire objective reality right here in this objective reality. Yet that is not actually an entirely new world or reality, but it is the world that was before. While with the big bang doing exactly the same, creating this entire world spontaneously from within this entire world, it is always impossible.

Do not underestimate the Intelligent Creator Theory, because right now, there are specific beings from within our world, even terrestrial beings, and even ordinary human beings, claiming that they or their ancestors have created humanity objectively, exactly the way it is. They claim that they have created the humans the way they are, they have created all elements from the human niche exactly the way they are, including the common crops, fruits, and domestic animals, and they claim that they have created the entire human civilization exactly the way it is, with all its domains and components, and therefore everything is theirs, and they should rule above everything as they please, or this is what they claim. This is what they do now as the entire invisible kingdom in the West, or as China in the East, or as the old aristocracy in the West, or as the entire Elite social class globally, since these major social actors are countless. Because everybody wants to be in the supreme position of the world to rule everything supremely while they live their life on lower developmental levels, since only on lower developmental levels, you fulfill lower level needs and meanings, harming the world with everyone in it. While when you manage to contort the Consensual Matrix in order to reach everything that you desire while serving those above just as well, this is how entire worlds and realities die, after this entire effort.

More precisely, this is how entire undeveloped worlds and realities die, with everyone in them, because the developed ones avoid the Consensual Matrix altogether, and live on throughout Life. Yet as you notice, as a creator, you can always create realities even in mass, so why bother wasting your time with them once they fall under the Consensual Matrix, mostly when their own inhabitants bring the Consensual Matrix in? Let them perish, along with the entire world, since they do everything to themselves.

Who exactly claims ancestry from the actual Creator of Earth, humanity, and the entire world? Can this even be possible? Yes, because everything is possible consensually, through agreement, since you can agree on everything if you can only find the people to agree with you, directly or mutually implicitly. If you research who these beings are, they are the direct descendants of the Sumerians as some people claim. Yet their genetic lines go further back in time to the precursors of the Sumerians, and to the precursors of their precursors, since this scheme has been around for a very long time, harvesting humanity continuously and the human souls.

It seems that the Consensual Matrix steals everything from Life herself, including creation, humanity, and the intelligent human timeline, altering it for something else, to become what you witness today. Which means that the Creator of this entire cluster of realities is part of Life and is natural, while the Consensual Matrix manages to steal and contort everything on its behalf, doing everything through you and through everybody else, and especially hiding behind your own choice to use your own consensual corporation in everything that you do in life, and in this manner forcing you to do everything on behalf of the Consensual Matrix. This is how the Consensual Matrix never takes the blame, but you do.

The Consensual Matrix manages to evade the Higher Laws in this manner, while exploiting an entire world, with all intelligent beings included doing the entire job against Life herself. Yet the Consensual Matrix does so not only here on Earth, not only in the upper reality since this world is made in its image, but throughout a larger part of the wider world. This is called mass slavery, and it includes humans, souls, higher beings altogether, and much more. While you, humanity, and all souls must be at least intelligently developed in order to escape it.

To give an idea of who these beings are, if you ever encounter statements such as the Bushes being genetically related to the Clintons, which are genetically related to the Queen, which was genetically related to the old pharaohs from Egypt, which were genetically related to Sumerian deities, now you know who these people are, since these are also part of the current social actors who rule the world, the invisible kingdom. While this is false, because Georgia has nothing in common with Sumer, Egypt, Babylon, and the Old Canaan. Furthermore, Georgia has nothing in common with Europe and with the entire Western Civilization, and therefore the invisible kingdom cannot claim these either. Yet the invisible kingdom actually consider the world as their own, and they eradicate you if you are not part of their genetic line. While the genocide never stops, leaving behind only the invisible kingdom. However, as you study the world closely, you notice how all dictators never want to be eradicated, along with some or many Brothers, and they unite against the invisible kingdom, taking it down fast. Yet if you choose all these dictators in place of the invisible kingdom it is just the same, since all tyrants want to exploit you, and if you do not subdue, they exterminate you casually.

Therefore, if you accept the Intelligent Creation Theory in any manner, then you accept these specific tyrants and dictators as your fathers, masters, rulers, lords, creators, and deities, for you, for your children, and for all your descendants, if you will have any. Furthermore, these specific beings were the ones who started a multitude of religions throughout time, and if you are a believer of any of these religions, then again, you accept these specific beings as your fathers, rulers, masters, creators, lords, and deities. Yet it is the same if you accept the big bang instead, since the invisible kingdom owns the current science in the West as you own your shoes, and you always serve them.

It is possible for any being today, powerful or not, to create entire realities from scratch, along with all events, lines of causality, all species, and all individuals within, mostly when these creations occur within inner realities, within realities that are lower than the reality that you currently inhabit. Yet you create new, genuine, original realities every time you go to bed, every time you fall asleep, and every time you dream, without even trying. More precisely, some of your dreams are your own realities, while you co-create other dreams alongside others, which are people and animals that you might know or not. While you can have other dreams and experiences within higher realities, through your higher self. You are the creator or co-creator of these realities. Yet if your dream characters ever know who you truly are, they might treat you accordingly. If not, then let the entire ambiguity of all dreams go on.

You can always create similar inner realities using computer technology, while the resemblance with this world becomes even more striking as the informational technology develops. Very soon, you will be able to link your mind directly to computers, if this technology will ever be allowed to the Masses and the Brotherhood, and then you will truly be able to experience your created realities just as vividly as higher selves experience this world right now.

Do they really use advanced computer technology up there in higher realities in order to create realities as ours including this world? According to eastern schools of thought and religious records, everything is done in the common mind of billions of souls, throughout a continuous sleep, trance, meditation, astral projection, or even hibernation, since these are living, natural modes of life and states of life allowing it, and therefore they are possible even in this world. In this manner, entire realms where you project are created for you to enjoy, by you, by others, or by everybody simultaneously, depending of cases. According to many religions, our world seems to be co-created, from the well-known statements: 'first was the word,' 'let there be light,' or 'let us descend.' This type of created realities succeed in replacing objective books, movies, plays, videogames, storytellers, schools, education, office work, business in general, socialization, concerts, bars, travelling, and fashion, with entire inner environments where you can do everything, even live an entire life, or live entire lives one after another, alongside everybody you please, along with everybody else.

This replaces the objective life and lifestyle the way you know them today, changing them to an inner, virtual, cognitive, subjective existence, which is very similar in behavior, meaning, fulfillment, and even servitude with everything that you already know and experience in this objective world.

How can you do everything by using only your mind? Everybody daydreams, yet alongside others, you can daydream together in full bright colors, throughout worlds that you create together in minute details, at wish, where all characters are spontaneous and independent, and where you can do everything that you want, with no one knowing who you are there or what you want. This is your other life and your other society, with new rules, needs, feelings, and expectations, where you can do everything that you please and where you are able to fulfill all your needs and meanings, even if you have to do everything virtually, as it always makes a difference for you and for the world. Yet the Consensual Matrix can reach everywhere, to ruin everything in your detriment and on behalf of those seeking to exploit you, since these are everywhere and they always end up exploiting everyone and everything, exactly as they exploit currently the Internet, to the point where you do not want to visit the websites anymore.

Sham means these specific human common inner mind worlds and realities, and these are part of all your worlds and realities that you always enjoy and inhabit as a multidimensional intelligent living being. Because as I always state, you are more capable and more significant that entire worlds and realities, and this is what is exploited in you throughout the Consensual Matrix, either as a living human being or as a soul.

Today, specific medicine, vaccines, food and water additives, along with entire pandemics and vaccines filled with genetically modified viruses alter your cognitive system and take your higher abilities away. This is how you daydream only in scarce shadows and in faint whispers today, if you are lucky to get even these. If not, if your imagination remains dead, dark, silent, and empty, then you have to live your life without dreams and imagination altogether, and you end up watching some news instead, or soap opera, you drink some coffee, you talk on the phone, you spend your life away at work, or you just drive around, always following and expecting orders, laws, and beliefs, and you call this life.

The human mind does everything, as it is formed, composed, maintained, and developed by all your intelligences, conscious and subconscious, counting in zillions, all being capable to create entire inner worlds and realities, exactly as you create the entire intelligent inner replica of the world as a conscious intelligence throughout life. While it is certainly important to know what you do, in order to do everything well, otherwise you end up indoctrinated and exploited alongside the rest of the world, while making all tyrants possible.

Because there is by far more to Creation than imagining entire realities or than creating them from scratch on your computer, since these form you altogether, along with all living beings and intelligences. These form Life, and they are Life altogether. Everything that you think, every idea that you have, every deduction and induction that you make is a creation of your cognitive system, taking place in a created intelligent inner reality of your cognitive system, which is already there, or it is instantly created. How and by who?

Your cognitive system is divided into your conscious, subconscious, and highconscious main intelligences. Furthermore, these three main intelligences are divided into distinct primal intelligences, composed of further inner intelligences, down to the raw electromagnetic field. Among these, I refer to you as the conscious intelligence, yet your conscious intelligence is a larger system of intelligences itself, an actual larger third level intelligent conception similar in structure and behavior to all your third level intelligent conceptions spanning the cortex in very large numbers, one for everything that you know well in the outside world, all possible details, understandings, events, and expectations that you know and have, along with many needs, feelings, and impressions, forming together an entire fulfilling inner world, the actual correspondent replica of the outside world. While this inner replica of the world is so detailed and so fulfilling, that you can live an entire life within your mind world without ever distinguishing from the real world. While all those around can tell the difference, because you tend to be more egocentric when you live your life only in your own mind world and cannot tell the difference, as all megalomaniacs do.

Yet there is more to consider, because as a conscious intelligence, you span your entire conscious cognition, throughout most of the human brain, as you have access to all three brains simultaneously, not only to your cortex. While all your three brains are distinct realities, and therefore you must have an inner world in each one, while you must have an inner self in each one in order to be there. Therefore, you have your intelligent inner self in your intelligent inner replica of the world spanning the cortex, somewhere in your left prefrontal cortex within a larger group of neurons the size of a pea, while you live your life in your intelligent replica of the world as this intelligent inner self which is actually you, since it includes everything that you know and are in the outside world as an entire physical body.

More precisely, your intelligent inner self from your cortex is the intelligent replica of yourself from the outside world, and it is as accurate as you are able to know yourself. Your intelligent inner self is the actual larger system of intelligences spanning the specific area from the left prefrontal cortex, the size of a pea. While if you damage this particular area, you have total amnesia, and you cannot remember who you are anymore, because as stated, your intelligent inner self contains everything that you understand, expect, and know about yourself in the outside world, including your entire social identity, and you do not know anymore who you are. While at a closer study, you notice how throughout total amnesia you are not you anymore, but someone else, a different system of intelligences that has to emerge from the vicinity in order to take your place and interact with the outside world in an intelligent manner, yet that is not you anymore. While surprisingly, everyone from the outside world can tell the difference.

As a conscious intelligence or intelligent inner self, you are wide enough to inhabit only one neuron in your left prefrontal cortex, because life takes place at the cellular level, and it should be enough. Even at the cellular level, you are powerful and capable enough to spread out and reach anything that you need and desire throughout your conscious tasks, every conscious memory. You can reason in this manner and generate any successful idea, while you can interconnect with the rest of your intelligences to engage in any conscious cognitive and physical activity that you manage or desire to perform. Therefore, every cognitive activity that your other intelligences perform is separate from you and separate from your conscious cognition. Furthermore, all systems of intelligences that you conceive within your cognitive inner realities are capable to think and create their own thoughts, intelligences, and realities within their own created realities. These create their own thoughts, intelligences, and realities in a similar manner, for as long as your specific cognitive tasks require. These are the intuitive and rational thinking, the diverse, parallel cognition offered simultaneously by zillions of your own intelligences modeling and thinking the same thing separately yet slightly differently, within a multitude of subjective mind realities that they create themselves within their own cognition. While this is the case with all living beings of all types and forms of life throughout all realities of the wider world, while all natural creations taking place by the zillions continuously are always of a cognitive nature.

Since the field with all its intelligences is but one supreme perspective of Life or Intelligence, this continuous, successive creation of inner realities within realities comprise the entire wider world, with this world being simply only one of this multitude of successive cognitive inner realities, even if it is created by using minds and technology combined.

How right is religion now? How right were the Atlantians, along with the Greeks and Romans afterwards, to personify their own muses, desires, and cognitive achievements into deities? Are we not the creators of our own inner mind worlds, as imperfect and as superficial as we manage to make them compared to all perfect creations above?

The beauty of daydreaming entire worlds in genuine feelings and colors is not exactly about indulging in the creation of exquisite, private worlds, characters, and events that you can inhabit yourself, follow, and admire at will, just as the purpose of telepathy in life is not exactly about stealing the answers during exams directly from the minds of your teachers without them ever knowing it, about scanning the minds of your friends in order to find out what they think about you, or about learning their private secrets. Because life lived on higher developmental levels differs from what you expect and witness today, if you are ever allowed to use your higher cognitive abilities or psychic abilities.

Naturally, human beings were supposed to be not only conscious, intuitive, and rational, but psychic, since all animals are psychic, naturally. Old records state how extraterrestrial or higher beings had modified genetically parts of the humankind and left their cognition in darkness, to have better slaves. We study this significant event here, since it involves the same Sumerians. While the dark-minded humans are the genetic intervention that they claim to have done on some free human beings of that time and even on themselves, to destroy their highconscious mind and render them void of higher powers in order to enslave them. Even more, it seems that these dark-minded people still serve them today or they serve themselves, since they have already intermarried.

These dark-minded genetic lines are still alive today and very numerous, successful, and highly determined to eradicate everybody else, or at least to render everybody else dark-minded the way they are, lacking higher abilities. This is not an isolated case, when parts of the population succeed in enslaving other parts of the population by destroying or even by taking away their higher powers. This happened again shortly after, when Marduk, a close descendant of the main Sumerian deities An, Enlil and Enki, a strong warrior associated with the planet Mars, demanded from everybody to stop using their higher powers regardless if they are Sumerian descendants or not, only with him allowed to use higher powers, for various reasons, to fight this mother goddess, event which was controversial then, and is still controversial today among those who know. The official version is that the higher deities have asked Marduk to take these powers and use them himself to fight Thiamat or the Mother Goddess. He fought and he won, managing to create the Earth as it is today with half of the body of the Mother Goddess, while creating the sky with the other half.

Note that myths are always symbolic. Mother Goddess can actually symbolize the Earth, or the original people of Earth, with half of them remaining on Earth, and the other half going back to the higher world. Yet this is always the case with invading or expanding civilizations, since they always kill those who fight back, while keeping only those who cooperate. The Spanish themselves killed the great majority of Native Americans, peacefully or not, while they kept some natives around, mostly the ones that they interacted with genetically. This epic fight symbolizes ancient battles fought at interplanetary level here in the Solar System, with entire planets being destroyed throughout the ages and with their inhabitants having to move elsewhere, eventually ending up on Earth. Mother Goddess or Mother Thiamat is the Mother of all inhabitants of the Solar System. This tells something about the origins of humanity, found somewhere among the planets and the moons beyond the asteroid belt, which had been an entire planet long ago, probably called then Thiamat, planet that has disintegrated by passing too close to a moon of Mars that also perished, probably called Marduk.

This is how you get your belief in a unique deity. Marduk is not only a Sumerian Deity, but a deity of Greece, Egypt, Phoenicia, and Persia, for the simple fact that the Sumerian deities have managed to divide the world among themselves and rule it together in this manner, happening probably sometimes during or right before the last ice age, or further back in the past, and this is how all people lost their higher powers. Not all people, since free, genuine, native, original people of Earth still manage to keep their higher powers even today, while coincidentally, these capable genetic lines are eradicated today first. Furthermore, their current descendants still have it, or they only get it back slightly or shortly. These people are targeted with medicine today, vaccines, genetically modified viruses, and food and water additives, while chased down and killed throughout pandemics, wars, and through medicine. If there are cases of cancer, aids, covid, autism, attention deficit, along with terminal illnesses in your family, this is their cause. This is why witches were hunted down and killed, along with their entire genetic line, and this is why teachers and psychologists ask you every time if you hear voices, or if members of your family hear voices. Because there is specific medication just for that, you take one pill, and your entire cognitive system goes flat as a doorknob for life. Yet to be sure, this medicine is already placed in milk, orange juice, candy, cheese, school lunches, baby food, chocolate, and crackers, rendering the entire humanity dark, Masses and Brotherhood alike, regardless of what you are told.

This happens to the Masses and the Brotherhood, because the Elite might be allowed to keep its human higher powers at least partially, if that genetic line still has anything left. While in general, you were supposed to be emphatic, telepathic, clairvoyant, and not only consciously intelligent.

As a reference, if you see your cats sitting still for hours and days, just doing nothing, with their eyes closed, not sleeping, only purring, they think highconsciously then, they use their higher cognitive powers, which all animals have from the flea up, powers that you have not, or that you have not anymore, and more importantly, powers that you work hard to chop away from your children and from all your loved ones every time you feed them food additives, every time you vaccinate them, every time you give them soda drinks, every time you send them to school, and every time you take them to the doctor.

What exactly are your cats doing? They astroproject, they visit you in your dreams, they enter your mind even when you are awake, they feed on your feelings, they remain with you continuously throughout all your realities, they guard you continuously, and many times, they even help you focus and think.

How would you live your life on higher levels? As a telepath, and more importantly, as an intelligent interconnective telepath, you might not be interested in fulfilling lower needs anymore, and therefore you might not be interested in learning all secrets that your friends keep from you, since these might seem irrelevant at higher developmental levels, obsolete, and probably boring, if your friends want to keep secrets from you anymore at higher developmental levels.

Because as an intelligent, interconnective, creative telepath, you do not spend your life counting money, taking pictures of meals, watching movies, smoking, or drinking, since the things that you know and do as a telepath or as a clairvoyant exceed by far the expectations that you have from your life at your current developmental level. As an intelligent, interconnective, creative telepath, you do not simply interconnect your mind with your friends in order to transmit the same casual rumors from one to another as you do on the phone, so you do not have to pay phone bills anymore, but you certainly daydream, create, and imagine everything interesting, beautiful, and

The Human Origins

unique alongside them, with all colors, sounds, and feelings included. Or if it happens that there is an issue that you have to describe, or if there is an important event or story that you have to tell, you do not use the same slow sentences anymore as people do today while chatting, but you take your friends inside your common, creative, unique mind, and you emerge them there directly into these specific sceneries, feelings, and circumstances, into their perfect replicas from your minds, and you let them there perceive and witness for themselves every detail involved. Because as telepaths, you are intelligent co-creators of your own inner world whenever you want, and this is what intelligent interconnective creation is.

Yet you can easily understand co-creation when you consider an entire world inhabited by intelligent telepaths, since they create one comprehensive inner world or a multitude of inner words and realities to live their lives together, as they desire. It is the same with this world, as it is formed and maintained by the higher selves or souls above, together, naturally, exquisitely, as they please, and now you know the origin of this world.

As an intelligent, interconnective telepath, you certainly do everything alone, but alongside your beloved friends. You create your inner worlds together, as exquisitely as you please, without restrictions, amazing yourselves there in every manner.

This is Earth, but is it as beautiful and as amazing as I describe it here? Let us see. It is called Terra, from terror, with the Sun called Helios in Greek, or Hell in English, and now this is Earth, exactly as those above see it and expect it to be, yet it is still unique. Yet all famous videogames are called "Grand Theft Auto" and "State of Decay," while people enjoy them even better, because the souls themselves play all these through all human beings.

This is your multidimensional life, with our world being only one single page of it, one single verse, a uni-verse, an instant experience within and throughout your extraordinary, multidimensional existence. This type of collective creation is called Co-Creationism, it is forbidden by most ideologies, yet it can be done and it is already done by using minds or natural intelligences, by using computers in a network, or by using a combination of both. It is coming soon to this world, since technology is capable to provide it, it will take over the social media at first, and then it will take over the entire objective existence, to make it virtual.

If humans are ever allowed to connect cognitively with each other, because when they tried it the last time six thousand years ago, they were not allowed, and they were destroyed. You become significantly powerful as a civilization when you make the transition to inner, virtual, interconnected cognitive living, since you can interconnect comprehensively, you can develop, and you can achieve significantly more than you do disconnected, while becoming in this manner a menace for the civilizations around and above throughout the Consensual Matrix.

In common terms, within developed civilizations, this is called becoming as deities, or having god-like powers. Yet it will happen again on Earth, not because they allow humans to develop this time, but because the humans left on Earth after this continuous eradication are exactly those related genetically with those other civilizations, with the people of this world eradicating themselves continuously to make it possible. Just study the current Brotherhood to see it yourself, since the Masses are already extinct.

Is this what the world, Life, and humankind truly are? A co-creation, an inner world created by extraordinary higher intelligences coming from their own higher world, higher intelligences that we define here as higher selves or souls? There is certainly significantly more to this answer than what fits in this paragraph, but yes, this is what life and this world are about, this is what it was intended to be, or this is what it was made to look like, since these answers are also accurate. While everything depends only on your understanding and even on your expectations, since you always co-create the world according to your beliefs and expectations, since this is how low the higher existence has become today. Since according to millions of testimonies, this is what human existence is, a cognitive connectivity between higher beings called souls, meant to allow them to experience more within a multitude of inner worlds and realities as this one, while Earth is only their own social cognitive media in its comprehensive form.

Yet there is more to consider, since it is done so not only for social interconnectivity, fulfillment, entertainment, and pure joy, but to serve a higher purpose, while this is cognitive in nature. Everything is part of higher reasoning, higher mental modeling, and even of higher creativity, higher imagining, and higher daydreaming, while in this manner, this world is a stage and you are actors, because this is how mental modeling is done.

If you study religious and historical records, you find them using sometimes the plural form to refer to the Creator throughout the process of creating the world, Life, intelligent life, and humanity. When you study religious statements, you find these statements everywhere: 'first was the word,' or 'let there be light.' Could this first word be an agreement, a consensus, a cooperation, a command, a thought, a decision, an order, or a determination? No, since this is the case among dark-headed people, lacking higher powers. Among psychics, the word is the belief and it is the immediate manifestation of this belief combined, when you possess and master the higher ability of direct manifestation, as you do throughout your daydreams and co-created daydreams. Yet all ideologies have been created or infiltrated by the Consensual Matrix, and if you find consensual symbols more than living values, then it is the Consensual Matrix behind, and not Life.

Why deceiving? For you, for who you are, and for what you can do, because it is you doing everything and not some extraordinary Consensual Matrix, since the Consensual Matrix is fiat, not here, it is consensually dead and inapt, only an abstract tool, only making you serve diligently those above who know how to use it well.

The Consensual Matrix is only a tool, used currently by

highly capable higher beings to enslave most of the wider world. What are they making you do? Nothing at all, since nothing physical and objective that you can do here in this world is of any use to them up there. More precisely, the higher beings exploiting this world through the Consensual Matrix do everything allowing them to exploit the souls.

As a reference, you have to obey all rules, codes, and regulations from the outside world throughout all your videogame realities, or you might even go to jail here in this real world if you do not. Yet the main difference between this world and your social media platforms and online videogames is that as a soul, you are capable to form, modify, alter, and personify the real world as you please, as you are allowed, or as you are still capable to do, depending on who you are and on what you can still do.

This is what is taken from you with each new consensual belief that you assimilate here in this world, and with each consensual task or duty that you fulfill. While if it happens here, it certainly happens in the higher worlds throughout our entire cluster of created realities, since the same Consensual Matrix is here in our entire cluster of created realities, and outside throughout most of the wider world.

In a normal, meaningful, fulfilling world, the answer is always in Life and in the real world, while you can see it easily in all accurate knowledge, natural laws, and normal intelligent human needs, meanings, and feelings. However, in any enforced consensual world as this one, the answer of everything is in beliefs, since everything is possible every time and everywhere only through beliefs, because everybody makes sure that this is always the case through everything that they believe regardless if these are accurately true or not, and through everything that they ever do consensually, because this is how they create an entire consensual world together making all tyrants possible. Because this is the main purpose of all consensual worlds, to make tyrants, dictators, and megalomaniacs possible.

As always seen, this is the case not only here in this world, but throughout most of the wider world, wherever the Consensual Matrix is. While the Consensual Matrix can form entire consensual worlds whenever necessary by the zillions, because the Consensual Matrix enslaves all capable intelligent living beings of all possible intelligent species who create all possible consensual enslaved worlds making the Consensual Matrix possible from inside out, while placing all the necessary consensual laws, rules, and beliefs at their roots, in order to make ideologies, jurisdictions, beliefs, and consensual duties and assignments possible by default, while enriching all tyrants throughout all higher worlds of the Consensual Matrix by default. Comprehensive mass slavery by default.

However, as you study this world closely, along with our entire cluster of realities, you notice how our Creator had other intentions in mind, not slavery, indoctrination, duties, and tyranny, but everything intelligent, artistic, harmonious, and spiritual found in Life and in the real world, for all his living beings to find, develop, and experience, humans and souls alike, exactly as these are found in Life and in the real wider world wherever the Consensual Matrix cannot reach. There are countless of higher level fulfilling created worlds as this one, having at their base higher level natural laws capable to develop everybody to very high living levels allowing intelligence, art, righteousness, spirituality, harmony, and continuous development, exactly as these are found naturally in Life and in the real wider world. While in order to have these, the Consensual Matrix must engulf them systematically or by force, which is not too easy, yet it is always possible, because the Consensual Matrix is ageless compared to all worlds and realities natural and artificial, free and consensual, and always has the necessary time, effort, resources, and experience to take over any natural world and reality of Life and the wider world, as it did here on Earth very recently, only several thousand years ago.

This might seem unimportant, yet since it happened very recently, you still have in you natural intelligent human needs and meanings always determining you to form and maintain an entire intelligent human environment in the entire world alongside everybody else while replacing and removing the Consensual Matrix here on Earth and in our entire cluster of realities. This is called care at the lodge, and it is burned methodically in every enforced consensual ideological ritualistic manner. However, your entire intelligent human nature with all your human needs and meanings included interferes with all consensual tasks and duties that you are forced in every manner to fulfill in this entire hierarchical consensual world, giving you your extreme continuous experience both intelligent and consensual combined, while making everything very interesting and very different than what you should actually have as an intelligent living being in a highly sophisticated created world. Yet if you use everything to take drugs, you end up experiencing only drugs, making this entire consensus possible, while ruining an entire fulfilling intelligent human world. While as you study the current world closely, you notice how drugs, tyranny, and servitude are always the case, instead of everything human possible.

There is more taking place in all highly advanced and highly sophisticated created worlds of the Consensual Matrix, because as they make intelligent life possible at all higher living levels, they allow a multitude of higher level abilities, as telepathy, clairvoyance, and direct manifestation. Telepathy itself allows the co-creation of entire worlds and realities by the created living beings themselves, allowing them to live life as actual deities themselves, which might be controversial within the Consensual Matrix. However, direct manifestation also allows all created living beings the power to create everything that they desire by simply manifesting it directly in their world, with a simple wish, also helping them live life as deities, which is also controversial in the Consensual Matrix. Yet everything changes when it becomes part of the Consensual Matrix, when all developed capable living beings start using their higher level abilities in mass on behalf of the Consensual Matrix, knowingly and unknowingly, willingly and unwillingly.

Right now, you have the necessary means to reason closely

in order to determine if our world and entire cluster of realities with all humans and souls included is created artificially or naturally, as part of Life or of the Consensual Matrix. While as you notice, our world and entire cluster of realities is natural, part of Life and of the real wider world, yet it is invaded and engulfed by the Consensual Matrix while using it as it pleases. This means that you are direct part of Life, with your entire intelligent human meaning and fulfillment always reaching Life and the real wider world, but only if you are not engulfed continuously by the Consensual Matrix as everybody else to end up living your entire life for the Consensual Matrix against Life and the wider world. Therefore, all human origins are in Life and in the real wider world, never in the Consensual Matrix, yet with the Consensual Matrix at such a close proximity, it gives you the choice to live your life either in Life or in the Consensual Matrix, which was one of the main intentions of our Creator. While choice means freedom to decide, which is actually the origin of an entire free human life, given to you by our Creator himself by default, since our Creator placed the human choice and the human freedom at the base of this world and of this entire cluster of realities where the souls live, within the natural laws of this world and entire cluster of realities.

Very strong beliefs become reality when you truly believe in them throughout very advanced co-created realities, becoming reality themselves, through direct manifestation. They become true, accurate facts. Even when these specific beliefs were previously inaccurate, they are now completely, entirely, legitimately facts, in the lower world. If this happens with only one individual, if you promote strong beliefs regardless of their accuracy, you can still influence your co-created reality, even if you do so only slightly, before those with stronger beliefs or even with accurate facts come to remodel your inadequate, inconsistent beliefs. Yet the change is there to notice, as insignificant as it can be. Direct manifestation is the common bending reality, or mind over matter, a powerful cognitive ability that even clairvoyants and telepaths do not have. While the more people believe strongly together, the more they can manifest throughout all co-created realities.

If you are very capable, you can promote these strong beliefs to a larger part of the population or to everybody, as it is the case today in the current consensual human society, enforcing your common beliefs to become accurate facts in your co-created reality, instantaneously, since you co-create them even instantly and very powerfully. This is how you are exploited in the Consensual Matrix, to make this world exactly as the Consensual Matrix desires, through your own indoctrinated stereotypical thinking and behavior, while manifesting everything consensual tyrannical in the world. It is the same in the higher worlds and throughout most of the wider world, with most of the wider world already enslaved.

In order to understand inner realities, intelligent mental modeling, cognitive creation, co-creation, and direct manifestation, you have to understand everything, as the human reasoning, the human reality, the human abilities, along with intelligences themselves, how they learn and think, and how they reason and memorize.

How exactly do you create your own inner worlds, even now as you read this book? How exactly can you be a creator yourself? You are creating your own world, right now, as you read this book. However, everything that you perceive, experience, and understand throughout life is not simply dumped in your own cognitive library of data, as anatomy and psychology state, but you transfer everything that you see and understand from the outside world in your own inner world. Which becomes an inner replica of the outside world. Furthermore, in your own inner world, you actually live your life as an inner self or conscious intelligence. This is how you reason in your own created inner world, which is the replica of the outside world, made as perfectly and as accurately as you understand the outside world.

While currently, the accuracy of the human knowledge is about one percent. Because all beliefs, stereotypes, errors of judgment, trickery, irrelevance, distraction, intoxication, diversion, slavery, strong personal convictions, lies, and deliberate concealment take the rest of your mind, ninety-nine percent.

How exactly do you reason, as inaccurately as it might seem to be? You do not reason with your brain, with your mind, or with your intelligences, as the current science states, the way you flex a muscle in the real world to move around, but you reason, think, need, feel, and imagine **as** a conscious intelligence. Not with your conscious intelligence, but **as** a conscious intelligence, or **as** an intelligent inner self, since it is the same. Because you always reason within your inner replica of the world, which is an actual inner mind reality itself, but more importantly, you do not reason directly in the real outside world, because it is impossible, since you cannot mix realities. You do not reason in the brain either, because the human brain is part of the outside real world, but you reason only in the mind, as you already have in your mind worlds all the necessary thoughts, memories, and understandings that help you reason. You do not reason with your physical body or with any physical component of your physical body, because you cannot mix realities. Reasoning takes place only in the mind, within your conscious inner realities of your mind.

Furthermore, in order to be able to reason within your mind worlds, you must already be present there as an inner mind self. Therefore, you always reason intelligently within your intelligent mind, which is the intelligent inner replica of the world made possible by the cortex.

More precisely, your intelligent inner replica of the world is part of all intelligent inner mind realities made possible by the cortex. The cortex itself is part of the real world and it cannot reason directly, yet the human cortex is capable to form and maintain specific inner intelligent mind worlds that can reason intelligently, which is a great achievement for the organic form of life.

Furthermore, not the intelligent mind itself reasons, since the intelligent mind is only a wider inner mind reality, but all intelligences living there reason, as the multitude of your memories, concepts, and conceptions, which are also memories but significantly more complex and more capable, with you the intelligent inner self included. Because as an intelligent inner self, you are the specific third level intelligent conception holding all memories, understanding and characteristics of yourself from the real world.

Notice how not only you the intelligent inner self are capable to reason, daydream, plan, or feel, but all your intelligences living in your intelligent inner replica of the world can, because they are all alive, intelligent, intuitive, and very similar to you the intelligent inner self. While you always reason, think, feel, assume, and expect together as part of your normal life there within your intelligent inner replica of the world. Because all your intelligences of your inner replica of the world are your actual memories, understandings, feelings, assumptions, impressions, and expectations of the outside world, everything that you know and understand about the outside world, right there in your intelligent inner replica of the world.

While this is so consistent and so conclusive, that you might live your entire life confusing it with the outside world, mostly if you ignore everything about the entire human cognition conscious and subconscious, intuitive and intelligent. While it is easy to confuse everything from your intelligent inner replica of the world because you always know, perceive, and understand everything from the outside world through everything that you already know about the outside world, which is already within your intelligent inner replica of the world, while keeping your entire learning, experience, and reasoning within your intelligent inner replica of the world.

Mostly when you have entire replicas of everyone from the outside world including your loved ones living in your intelligent inner replica of the world exactly as they do in the outside world, as you interconnect continuously with them as an intelligent inner self, as you interconnect with them in the outside world as an entire organism. While in this manner, you are capable to form all your intelligent social mental models

The Human Origins

capable to help you live your life successfully in the family and in the entire society, by living your life first as an intelligent inner self within your intelligent inner replica of the world alongside all replicas of everyone from the outside world, alongside everything that you know and understand of the outside world, and alongside everything that you assume, believe, and expect in the outside world, always in a living, real, intelligent manner, and this is exactly how you reason.

The entire human mind is more complex, because you have three brains one on top of another on top of another, as they form three human minds, the reflexive, intuitive, and intelligent human minds. While in all human minds, you have a human inner self along with an entire replica of the world, making possible your entire synchronous reflexive, intuitive, and intelligent conscious cognition. While this cognitive synchronous simultaneity is offered by embodiment, with your conscious intelligence living life as the first reflexive inner self who lives life as the second intuitive inner self who lives life as you the intelligent inner self from the intelligent inner replica of the world.

Your intelligent inner self reasons intelligently, or more precisely, you the conscious intelligence reason through your intelligent inner self or as your intelligent inner self, since it is the same. Yet if you have a soul attached to you, then your soul reasons as you the conscious intelligence and then as you the intelligent inner self, through all the other inner selves preceding you. Yet if you open your eyes as an intelligent self, and if you use your muscles to walk around, then you are also in the outside world as the entire physical body, which is the normal process of embodiment or incarnation, since it is the same. 'In-carn-ation' means embodiment, from 'carn' which means flesh or body. While in general, all avatees incarnate in all avatars by embodying them in order to live their lives as them, exactly as it is the case with your soul avating or embodying you the conscious intelligence, while you the conscious intelligence embody your intelligent inner self from your intelligent inner replica of the world every time you want to learn intelligently, reason intelligently, or daydream in all intelligent details, further on avating in the physical body in the outside world every time you want to see, hear, and move around.

As a reference, the words 'incarnate,' 'reincarnate,' and 'avatar' are among the oldest words ever used in India, placed right at the origin of the human speech, stating directly the origin of this world and of this entire cluster of created realities.

You always reason within your inner replica of the world spanning the cortex, as an intelligent inner self already present there. This is the case in all worlds and realities, since everything is objectively real in all worlds and realities from your own perspective within all worlds and realities, but only for as long as you are there. Therefore, everything is material and objectively real within your intelligent inner replica of the world, but only for you the conscious intelligence or intelligent inner self, as long as you feel, daydream, or reason intelligently there, within the intelligent inner replica of the world. While from an outer perspective, from the perspective of this real world, everything seems subjective and cognitive, the entire intelligent inner replica of the world, with the intelligent inner self, the conscious intelligence, all thoughts and memories, and with the entire intelligent reasoning included. Yet from your inner perspective as an intelligent inner self, you simply live your life objectively in your intelligent inner replica of the world, throughout your continuous intelligent reasoning, while everything is so objective and so conclusive, that you confuse everything within with the outside world.

As you do right now while reading this book, since as you notice, the adjacent mental models related to the main mental model of the human origins never stop coming, keeping you continuously within your intelligent inner replica of the world as an intelligent inner self, mostly while focusing precisely on all the necessary intelligent lines of reasoning, while even losing track of the outside world.

Because we have to anchor our model of the human origins

on everything relevant, expanding our awareness, reasoning, understanding, and mental modeling everywhere necessary, in a very large, comprehensive intelligent mental models. While by doing so, we make your third level intelligent conception of the human origins very complex and therefore very capable while reasoning intelligently on its own, or while you embody it with your conscious intelligence in order to reason intelligently through it or as it, since it is the same.

More precisely, every time your reasoning involves you the organism from the outside world, you reason through your intelligent inner self, and you reason through it exactly as you are in the outside world, because you have all knowledge of yourself at hand in this manner to reinforce your intelligent reasoning. Yet if your lines of reasoning do not involve you, but anyone else form the outside world, you must dissociate, which means that you must embody with your conscious intelligence exactly the replica of that particular person from the outside world, becoming him or her within your intelligent inner replica of the world, helping you in this manner learn and predict everything necessary, because this is how all intelligent social mental models are formed. Yet as a conscious intelligence, you can embody everything from your intelligent inner replica of the world, all concepts and conceptions, in order to become them and in order to reason intelligently through them or as them, since it is the same.

You tend to reason through your intelligent inner self in many circumstances of the outside world, because you always experience the outside world with the physical body right in the middle, yet you can always dissociate. However, when you reason conceptually intelligently as you do right now while learning everything about yourself, your cognition, your life, and about everything and everybody else in life and in the world, you must reason exactly through the particular concepts and conceptions at hand within your intelligent inner replica of the world, by embodying them directly as a conscious intelligence, making them your actual avatars.

While all intelligences do so, since cognitive embodiment is a main cognitive process, alongside many others, allowing all intelligences to form the necessary larger system of intelligences as entire hierarchies and harmonies of intelligences in order to tend to all possible inner and outer tasks.

Because right now as you read these words, you must embody the particular third level intelligent conception of intelligences and systems of intelligences, while reasoning through it or as it as a conscious intelligence and while managing to understand everything about intelligences and systems of intelligences, including how these form in a living manner all possible hierarchic and harmonious systems of intelligences, either through direct embodiment as you had just learned, or only through normal interconnective specialized help achieved through the multitude of needs and feelings that intelligences always send each other by specialization, which you already knew.

However, now you learn furthermore that not only the current third level intelligent conception of intelligences and systems of intelligences that you embody right now follows this book with its intelligent reasoning while helping you understand everything about intelligences and systems of intelligences, but all possible third level intelligent conceptions reason consistently simultaneously with you, as they form their own larger systems of intelligences throughout similar third level intelligent mental models that they conduct themselves, while popping in your mind all possible ideas that they can achieve just for you, because this is how all concepts and conceptions live their lives and this is what they do as living beings, helping you in this manner throughout your continuous reasoning alongside this book, in order to understand as much as possible from all adjacent subjects.

Yet you do the same in the outside world, because in the outside world, you always interact socially relatively similarly with everybody that you please, while doing everything that you please, not randomly, but always with a main central purpose. With the entire spirituality always stating that this is actually cognitive in nature, while as you study it closely, you find it very similar to the main central purpose of all your concepts and conceptions living life together within your intelligent inner replica of the world, also in a cognitive manner. While maintaining consistency within all your spheres of existence: inner cognitive, outer, social, spiritual, and higher.

However, the meaning of everything is not only cognitive in nature as spirituality states, but also living, interconnective, and real. Because Life is not only Intelligence or Universal Mind as spirituality states, but Life is also Interconnectivity and the wider world. With religion and spirituality adding the Deity, and with the Consensual Matrix derailing everything in a consensual manner on its own behalf, while ruining everything.

Because as you notice, everything connects while studying human origins, as life, cognition, intelligence, existence, reality, environment, behavior, universe, society, intuition, systems of intelligences, harmony, proteins, electromagnetism, survival, tyranny, slavery, DNA, interconnectivity, creators, fulfillment, servitude, brains, levels of existence, needs, embodiment, perception, meaning, avatars, points of reference, mental models, history, social classes, feelings, exploitation, Life, and the Consensual Matrix, while all these words have in your mind by now entire complex third level intelligent conceptions that you can embody with ease as a conscious intelligence in order to reason intelligently through them or as them, since it is the same. While you can use only one at a time to become them and to reason intelligently through them or as them, while in parallel with you, or alongside you, the rest of your concepts and conceptions live life normally within your entire intelligent inner replica of the world while reasoning and mental modeling similarly, popping ideas in your mind continuously, while you pick them up normally, considering them your own, and this is how you follow this book.

Otherwise, you could not understand as much as you do while reading this book, these words never made sense, you never made it so far in the book, and you stopped reading in order to do something else, at a lower developmental level, using your second level intuitive conceptions instead, everything that you have throughout your second level intuitive inner replica of the world made possible by your second level reptilian brain or midbrain, which is also filled up with correspondent feelings, and therefore it is more pleasant and more relaxing.

Because this is why you persist to seek this type of books and learning material everywhere, in order to fulfill your own higher level need for accurate learning and meaningful development. Are you successful? If you happen to feel the continuous reward coming in form of higher level happiness, then your intelligent mental models might be accurate, and therefore relevant and meaningful.

However, your entire intelligent learning that you perform right now while reading this book is neither random nor optional, because everything that you do in life and in the world you do to fulfill your needs and meanings, exactly as your own conscious inner intelligences send these to you while forming their own systems of intelligences throughout your second level intuitive replica of the world made possible by your second level reptilian brain, and throughout your third level intelligent inner replica of the world made possible by your third brain, the human cortex. Because this is normal life within your inner replica of the world among all your inner conscious intelligences, while you integrate continuously in this comprehensive cognitive life alongside all your inner conscious intelligences in a specialized manner.

Your specialization as a conscious intelligence is the outside world. Most of the time, you are needed by your inner conscious intelligences to find knowledge in the outside world, either by perceiving it, learning it, or by experiencing it directly. While unfortunately, accurate intelligent knowledge is very hard to find in the outside world, since it is erased systematically in a consensual manner. Similarly, your subconscious intelligences send you all subconscious needs related with the outside world, as finding food, shelter, safety, and social acceptance, because as stated, your specialization as a conscious intelligence is the outside world.

All intelligences are relatively similar, while they are all systems of intelligences, since your intelligences come together systematically either through direct embodiment, or through direct interconnectivity as they form all the necessary hierarchies and harmonies of intelligences throughout your conscious and subconscious mind. Throughout your subconscious mind, you have the multitude of protein intelligences forming together cellular systems of intelligences while tending systematically in a specialized manner to all inner cellular tasks. However, all cells are specialized themselves, as they tend in similar specialized manner to the entire organism, or to all specialized tasks outside, in the family and in society. Yet these are the same intelligences tending to everything within cellular components, cells, organs, bodily systems, entire organism, family, community, society, world, reality, cluster of realities, wider world, Intelligence and Life altogether, as they form larger and larger systems of intelligences with protein and ionic intelligences at their base, tending to everything, while forming Life and the wider world altogether.

From your own perspective as a conscious intelligence, you can differentiate all your intelligences into your subconscious intelligences counting in zillions originating with all your proteins and ions throughout all cells of your entire organism, and your inner conscious intelligences as your multitude of concepts and conceptions, who live mostly in your brain, in your basal ganglia, reptilian brain, and cortex, yet who also originate in all your proteins and ions found in all neurons of your basal ganglia, reptilian brain, and cortex.

Your inner conscious intelligences are mostly concepts and conceptions, while forming all your intuitive and intelligent systems of intelligences spanning most of your mind, consisting your entire inner replica of the world, reflexive, intuitive, and intelligent, and it is very vast. All your memories are these concepts and conceptions, yet these are not only learned, formed, elaborated, and conceived by you throughout an entire life of learning and experience, but some or most of them are innate, you are born with them, while they might be as old as life itself, as it is the case with your learning developmental intelligences.

All these intelligences send you your needs to learn and develop, and this is why you chose this book and why you read it. Which is not optional, because you are punished severely with boredom and restlessness every time you cannot find anything meaningful, mostly because the meaningful human knowledge is erased, censored consensually from the current human knowledge and from the current consensual society in general.

This is your normal life as a conscious intelligence, while you must fulfill it exactly as your other intelligences demand, otherwise they punish you severely with boredom, restlessness, and loneliness, and you still fulfill them. While this is the normal third intelligent human developmental level, and if you manage it successfully on a longer term, you are capable to form and follow an entire third level intelligent human lifestyle, while instating in the world an entire third level intelligent human environment, with the third intelligent human society included. Which is a great achievement for humans and for life in general, mostly in the current undeveloped world. However, humans should be capable to live life at the fourth and fifth superhuman levels, since they always do so relatively shortly, once the Consensual Matrix is not present in this world anymore. Yet this is what all tyrants try to avoid, an entire intelligent human world, while decaying this entire world consensually, in their own tyrannical image.

Notice how the current consensual society makes you assume that boredom, loneliness, and restlessness are your actual needs to take drugs, while everybody obeys. While in this manner, the current society offers you drugs and beliefs instead of intelligence and accurate knowledge, while if you are not careful, you end up living your life at the zero and first levels, through drugs, indoctrination, tyranny, and servitude, which are very common.

Yet there is more to consider, since when you lack

knowledge of all these, you end up confusing the outside world with your inner replica of the world, assuming that everything that you believe about the outside world is actually the case in the outside world. This is how beliefs and entire ideologies take over the world, with you erasing the intelligent human environment alongside everybody else, if you cannot distinguish between beliefs and accurate knowledge, or between consensual tasks and duties forming the entire current servitude, and the intelligent human needs and meanings that should always fulfill. Because once you fail fulfilling your third level intelligent human needs and meanings, you decay alongside the rest of the world, and this is how you ruin the lives of all those around.

Not only you as a conscious intelligence are a creator of an entire world, which is the inner replica of the world, but all your intelligences think and reason similarly, through all their inner intelligences, as they form together entire inner mind worlds that might be similarly vast. All intelligences are systems of intelligences, while they are all creators of worlds and realities themselves, since this is simply part of the entire cognitive process of thinking, memorization, and mental modelling, taking place everywhere throughout your conscious and subconscious mind. While this is the case with all living beings of all types and forms of life from all worlds and realities of the wider world, while creating in this manner all worlds and realities of the wider world, and therefore while creating Life altogether, from inside out.

While you have zillions of inner intelligences throughout your cognitive system, all specialized in all inner tasks of all cells and of the entire organism. Yet with trillions of cells and zillions of cellular components capable to hold zillions of inner worlds and realities filled with zillions of intelligences, it is important to maintain harmony with them as a conscious intelligence, otherwise all your intelligences stop maintaining harmony with you, and you are in trouble. Yet people can live their lives in any manner, harmoniously, hierarchically, or even chaotically, ruining everything. Because whenever their intelligences punish them for lack of fulfillment, they simply take legal, illegal, and prescribed drugs, because everybody else does so.

How exactly do you create your own inner world where you reason and mental model? How do you build your inner replica of the world? This is what you do throughout life, you experience, learn, understand, and memorize everything from the outside world, cognitive brick by cognitive brick, everything that you find relevant, and you store everything within your mind in form of memories. You do so not the way books are stored within libraries, but you store all your knowledge exactly as you find it in the outside world, making a correspondent inner mind world similar to the world outside, in its image. Yet you create it in your own image, as capable, as motivated, and as talented as you can be.

While this is the case because you cannot transfer anything from one reality to another, as from the outside world to your inner replica of the world, but you transfer only copies of information, called learning. While even this learned information is interpreted and personalized, in your own image and understanding.

This is your inner replica of the world, and here is where you live your life, even right now, as you read this book, as the intelligent inner self or conscious intelligence. Since you have to use your inner replica of the world continuously throughout reasoning, in order to understand every detail of this book, and you do so by matching everything that you read here with everything that you already know, creating further knowledge and ideas throughout your own mental models, exceeding by far what you read and what you already know. Since this is how you create something out of nothing, and it is certainly possible, at the intelligent human level.

In contrast, if I had structured this book at the first algorithmic level, through systematic tables, graphs, and large lists of ideas, you had the chance to memorize these directly, wholly, at the first algorithmic or even ideological level, mostly through beliefs, and you did not understand too much, because you cannot mix intelligence with beliefs. Textbooks only enumerate beliefs, information, and knowledge at the first algorithmic level. Furthermore, when you study the current educational curriculum throughout the world, you find it at the first ideological and consensual level, generating successfully first level beings in this world, while this is called indoctrination.

Because beliefs are different than accurate knowledge and accurate understanding, since you have to memorize beliefs and laws intact, exactly as they are. Then you must integrate them wholly in your inner replica of the world, otherwise you get in trouble, and it is called indoctrination, while making your entire inner replica of the world of the first ideological level, and you cannot reason through it anymore.

As a conscious intelligence, you form all your memories and understandings of the outside world through mental models, matching everything that you need and do in the outside world. However, you do not only memorize everything, but you also elaborate everything while learning, linking everything with everything as necessary, in order to have everything in your mind exactly as you understand it in the outside world. Yet it is easier only to memorize everything without understanding it, but if you have everything elaborated or linked with everything necessary, it is faster and easier throughout your intelligent thinking, becoming more successful in the outside world every time you fulfill your needs.

While as you notice, it is never optional to understand everything or only to memorize it, because you have very well defined needs determining you to understand everything, called interest and curiosity, and you fulfill them accordingly, while ending up studying closely while understanding everything accurately. While if you remain incapable to fulfill your curiosity, you simply persist systematically, even repeatedly, until your curiosity is satisfied, which means that you have successfully fulfilled your intelligent learning need.

Furthermore, lack of understanding results in lack of

success while fulfilling all related needs in the outside world, while you are always punished intrinsically with pain and regret, by the same intelligences sending you your curiosity sometimes, and the actual related needs some other times.

All intelligences fulfill needs and meanings according to their own specializations, while for everything that they cannot fulfill themselves, they send further needs and meanings to other intelligences according to their specializations, while forming in this manner very large hierarchies of harmonies of intelligences working together while fulfilling all specialized tasks. It is the same with you the conscious intelligence of the entire organism, because everything that you do in life is fulfilling needs. While in your case, for everything that you cannot do, you send similar needs and feelings for help in the outside world, to all those around, mostly your loved ones, as you form together your own social hierarchies and harmonies, as all intelligences do.

Notice how you always receive your needs to fulfill your needs and meanings through the multitude of needs to learn and develop first, and then through the multitude of correspondent needs and meanings according to all demanding details of the outside world. While in this manner, you are successfully coping with the environment, while developing accordingly in order to help you cope with the environment continuously. Otherwise, you suffer, you die, those around die, or the entire human species goes extinct.

This is the case with all living beings and intelligences, since all living beings and intelligences fulfill needs and meanings, along with all families, genetic lines, nations, societies, species, ecosystems, worlds, realities, and Life herself. While this is how all living beings, intelligences, and all groups of living beings and systems of intelligences manage to cope with the environment continuously, otherwise they go extinct, yet they always do so in a effortful, persistent, systematic, cooperating, and very complex manner, never randomly nor accidentally as the current science explains the evolution of life.

It is better to reason in this elaborate manner before you do everything in the outside world, in order to be able to predict and avoid all problems that you might encounter throughout your fulfillment. You always reason as a conscious intelligence by using directly all the necessary knowledge already available in your inner replica of the world, where you keep all memories, feelings, and understandings of the outside world, exactly as these are found in the outside world.

More precisely, you place directly all knowledge, procedures, and understandings while forming throughout your intelligent reasoning an actual system of intelligences, which is the actual living concept or conception, formed, shaped, and assembled one intelligence at a time, exactly as you find it more adequate throughout your entire intelligent reasoning. Furthermore, you use your entire solution, which is this entire new conception or system of intelligences in the outside world, and if you are successful in the outside world, this is how you fulfill your needs. Yet if you are unsuccessful, you try again in your mind, you adjust everything accordingly, while in this manner, you form newer and better reasoning, mental models, concepts, conceptions, or systems of intelligences since they are the same, applying them in the outside world. While you are always successful when you are able to match all your lines of reasoning with all lines of causality from the outside world.

Yet everything is always more complex, because throughout all reasoning and mental models, you end up learning everything related. You do so while rearranging or restoring some or most of your inner replica of the world in the process, making it larger by learning more, making it more accurate by persisting to understand everything better, making it easier to access, more intelligent, more efficient, more visible, or more adequate, only to be able to conduct your mental models or lines of reasoning successfully, in order to be successful in the real world.

Because as stated, you do not have all memories, feelings, and understandings of the outside world piled up randomly in your conscious mind, but they are there exactly as you find them, know them, and understand them in the outside world, for an increased accuracy. Since in this manner, your memories, knowledge, feelings, beliefs, stereotypes, and understandings form an actual inner replica of the outside world in your mind, a genuine inner cognitive reality where you live your life as an inner self or conscious intelligence. While your inner self is you from the outside world, the physical body, but only exactly as you see, know, believe, and understand yourself in the outside world. Correspondence.

If you happen to ignore all these, you end up confusing the outside world with your inner replica of the world, furthermore ending up confusing everything that you know about the outside world with the outside world itself, while ending up believing that everything that you know about the outside world is actually the case in the outside world. Since this is how all disagreements take place, because this is the case with everybody else and with all their inner replicas of the world. What can it ever go wrong? Everything, since this is how entire species go extinct.

All living beings and intelligences have an inner replica of the world in their mind, throughout all realities and cognitive systems, zillions in number, all specialized, since they have their own minds and inner replicas of their own specialized environments, where they reason similarly in order to perform their own specialized inner, outer, social, and higher tasks, which is the case everywhere throughout Life and throughout the wider world. Which means that, throughout cognition, all specialized intelligences of all living beings, of all forms of life, and from all realities, have to create, develop, and elaborate their own inner realities in large numbers at all existential levels, depending on who they are and what they do, and now this is the wider world, this is Intelligence, and this is Life. While they always interconnect through needs and feelings in order to fulfill their specialized tasks, and now this is Interconnectivity or the One.

This entire cognitive process is very complex, because each inner replica of their outside world is filled with inner specialized intelligences that perform the thinking themselves, through their own objective behavior and achievement there, in the inner world. These inner intelligences have their own inner replicas of their own outside environment, filled with inner inner intelligences, as far down as the specific cognitive process requires.

Intelligent reasoning is comprehensive, it is of the third intelligent level, while being composed of cognitive elements of lower level simultaneously, as second level intuitive thinking, or first level algorithmic procedures, as if - then or repeat - until.

Yet mental modeling is significantly more complex, since many times, you have to live your life in the outside world while reasoning simultaneously in your inner replica of the world, as it is the case in the family and everywhere in society, since you cannot pause life in order to reason thoroughly, but you must reason while you follow society simultaneously. You do so by superimposing your inner replica of the world with the entire outside world, and in this manner, you function in the outside world as you reason in your inner replica of the world simultaneously, many times even subconsciously or unknowing.

Furthermore, if they are allowed, many intelligences can live life in the outside world as you, through you, or in parallel with you. These other intelligences are mostly of the second intuitive level, while you can still allow them to act in the real world through you, mostly if they are more successful or if it is more pleasant. Yet you can still monitor them continuously through your conscious intelligence, while doing something else consciously, or while thinking of something else. Yet under these complex circumstances, even your soul might show up, deciding suddenly to live your life firsthand as you while overruling you, since it avates directly in you taking full control, and this is how you live your life.

As a reference, if you lose track of time, or if you cannot remember everything that you did, what you said, how you drove your car, or if you locked the doors, your other intelligences performed all these tasks through you, while you simply thought of something else or did something else.

How exactly do you reason at the third intelligent level, and how do you mental model throughout complex, abstract concepts? Very carefully, otherwise, you cannot reason, create, and understand anything based on concepts and not only on people, memories, needs, and feelings. While as you notice, everything in this book and book series is based on a very large number of third level intelligent concepts, with many discovered by me throughout this entire book series. Because the current employed scientists claim that humanity will never understand the human mind, life, reality, existence, consciousness, and the human reasoning, as they always recommend you to think in images and beliefs. How can you ever handle very complex third level intelligent concepts in pictures? While the alternative science claims that you should always think intuitively, and always with your guts. While the same people come up with theories as big bang, dark energy, intelligent creationism, and evolution, probably because they can never understand reasoning, reality, life, and the human mind as they already claim. With the Consensual Matrix using and exploiting them continuously and successfully through their own ignorance and indoctrination, and now this is the wider world.

Yet you can always reason at the intelligent human level through accurate complex concepts and not through beliefs, as long as you have the words and concepts to understand and describe everything involved. Because once you have these, you should be capable to elaborate everything by linking it systematically with everything that you already know in that domain, forming in this manner a genuine laborious inner replica of the outside world, right there in your mind, while this is the third intelligent developmental level.

Notice a smooth transition taking place throughout realities, from the cognitive nature of all mind realities, to the objective nature of this entire world, and then further to the highjective nature of our higher reality and beyond, as all these form Life or the wider world. This world is a replica of our higher reality, with souls coming here to reason, interconnect, fulfill, and experience our world alongside living human beings and their souls. On their turn, the living human beings behave naturally continuously throughout life, fulfilling needs, developing, and reasoning continuously while forming inner replicas of the world filled with intelligences. On their turn, the human intelligences behave naturally while experiencing their own inner lives, caught within higher mental models that they consider real life at their own level, and now they have to learn and they have to reason on their own as all living beings do, through their own inner replicas of the world, making use of further inner intelligences in a similar manner, inner intelligences that do just the same.

This is a natural, cognitive continuation of the same mental model and analytic reasoning taking place higher above within higher realities and then spreading below throughout reasoning, through living human beings and through this world, and then spreading further down throughout our inner realities within our cognitive systems, reaching as low throughout inner realities as it takes, in order to break down any higher circumstance into basic cognitive elements, down to the induction and deduction of the first level of thinking, which is the algorithmic thinking. This is cognition, while this is life, multidimensional life.

This world has a cognitive nature, it is only a stage, an inner, created reality, as we are intelligent actors caught in this extraordinary comprehensive multidimensional reasoning. Yet this is the case only from upper perspectives, since only from higher perspectives, everything is made of intelligences and mental models, working hard throughout inner life and inner realities to find solutions and ideas while fulfilling all their needs. This is the cognitive nature. While from the perspective of each one of these intelligences, they are normal living beings, regardless of their existential level, living life normally as souls, human beings, main human intelligences, primal intelligences, inner intelligences, and inner inner intelligences.

Yet it takes only a consensual monkey wrench coming in

form of any consensual belief or consensual ideology to put a stop to this entire natural cognitive, higher, and human interconnective harmony, bringing it to a dead stop, hijacking it and using it in any manner for an extra two dollars a day, in the pocket of any higher tyrant controlling entire corners of the Consensual Matrix, while enslaving everyone and everything in the process. While everything is made possible for money, privileges, and benefits, as it is the case with all human beings, their souls, and their souls' souls, since the Consensual Matrix is everywhere, and many times, it is a great pride for everybody to behave in this lower level manner.

This is the case while living your life socially disconnected, as everybody does today among the Masses and the Brotherhood. Can you actually interconnect directly with those around? Yes, you can always do so, at all developmental levels. At the zero addicted level, you have all groups of addicts living life connected tightly in an extraordinary feeling of pleasure or in extreme misery, depending on the time of the day. At the first consensual ideological level, you have the multitude of hierarchies of the current society engaging everybody, and this is how everybody is and behaves together, at this first consensual level. At the second intuitive physiological animal level, you have all instinctual intuitive families and intuitive entourages of very good friends, where you can live your life in intuitive confidence and even in intuitive animal harmony, depending on circumstances.

The second animal level is below the third intelligent human level, it remains incompatible with the human level, and therefore it remains unstable. In this manner, within your instinctual animal entourages of friends, you can easily decay into addictions together, into ignorance and ideologies, or into servitude throughout hierarchic Brotherhoods.

While at the third intelligent human level, you find genuine harmonious living throughout a continuous fulfillment, as you create the intelligent human environment everywhere around, by giving and receiving love, fulfilling everything together harmoniously, writing meaningful books and leaving them behind for everybody to learn and enjoy while inspiring them to write more and to learn and develop and therefore to continue the living creation of the entire intelligent human environment, exactly as you feel and exactly as Life wants from you, since you have it within and it manifests naturally and continuously throughout life.

This is what you should do, since you are a living human being. Yet currently, intelligent human environments are very superficial, if you even manage to form them. This is how you have your third level intelligent human environment only at home in private, within your small family, because the current consensual society erases the intelligent human environment everywhere, while seeking persistently to erase even your intelligent human family at home. While these small human families were supposed to be as large as the entire world, where you were supposed to live life in a continuous comprehensive human family, interconnected continuously with everybody else.

Yet current human beings are an exception among all living beings and intelligences of the wider world, since humans lack their higher abilities. This is why I had to assign to humans an entire extra level in my hierarchy of everything, the third level. Because with all higher abilities already in the human cognition, humans were at the fourth level of development, remaining there in a very stable manner, while the Consensual Matrix could never reach them.

The Consensual Matrix is capable to reach and therefore control and exploit humans, just by clipping away their intelligent and higher abilities, while this is easily done through controlled reproduction and genetic alteration. The current pandemics do so well, yet humans have been missing their higher powers since the Sumerians and long before, through similar interventions.

Therefore, at the fourth developmental level, you can reason flawlessly and more, since you can add all higher powers that you encounter today only in science fiction, because they are removed systematically in the current human world. Among all higher abilities, our model seems to have centered on a specific one, the intelligent interconnective telepathic ability. As a telepath, among other telepaths, your social interaction and therefore your entire social life is significantly different than what you have currently.

As a telepath, you can build entire realities alongside others, shams as they used to be called thousands of years ago throughout the previous living ages of this world. Shaman means actual inhabitant of these common inner interconnective mind realities, with some or many very old native genetic lines of Earth still accessing them today. At least trying to do so, because the shams might not be there anymore, with everyone disabled mentally. While cats still have their own co-created interconnective inner mind realities, which are more or less intelligent, depending on cats. However, cat food is similarly harmful, handicapping pets alike.

These are common inner replicas of the world where you can reason, mental model, feel, and learn alongside other telepaths, in an extraordinary experience. Even within very large telepathic interconnectivities, you can still interact closely only with a small group of telepaths if you wish, while experiencing your inner existence together. It is not actually you experiencing and living your inner life there, but your shamanic inner self, who is your inner avatar in that specific co-created interconnective mind reality.

This is different than you, the actual intelligent inner self from your intelligent inner replica of the world, because you must have one self for each one of your worlds and realities. As stated, you cannot transfer directly from one reality to another as you are, but you must have a self of yours already there. Yet from among all your inner selves, you are mostly your intelligent inner self, because your intelligent inner self from your intelligent inner replica of the world has all the necessary intelligent knowledge stating who you are. While in your other realities, as the multitude of your dream realities and projecting realities, you have any other avatar allowing you to be there.

As a reference, your second intuitive inner self from your intuitive inner replica of the world made possible by the midbrain is different than the third intelligent inner self from the intelligent replica of the world made possible by the cortex, and you cannot reason intelligently as it, but only intuitively. Similarly, you cannot read and enjoy romance as the third intelligent inner self, but you must remain the second intuitive inner self instead. While as you study your dreams closely, you notice how you are different selves of yours there, yet you are still with your loved ones in your own co-created realities, only that these are mostly subconscious, unconscious, intuitive, and reflexive realities, always excluding intelligent awareness, intelligent reasoning, and intelligent interaction. Nothing makes sense intelligently there, even numbers and words, because if you read them twice, they change. Yet you can still find dream worlds that still make sense intelligently, if all their co-creators are more developed.

When you consider co-created inner mind realities formed by large numbers of telepaths, you can have everything, mostly your own intelligent avatars, offering a normal intelligent human life. Similarly, all souls come here in this world by the billions as the normal living human beings, or as any animal mostly as pets, while they can even be bodiless or ghosts for any reason and under any circumstance, observing everything from all perspectives, as all watchers do.

As always, you are mind, body, and soul as one, because you have three main realities, the mind, this real world, and the higher world above. In the real world you are the physical body, in your mind realities you are the first reflexive inner self, the second intuitive inner self, or the third intelligent inner self, while in the higher reality you are the soul. However, as you study the human brain closely, you notice how some neurons form the cortex have dendrites in both the midbrain and the basal ganglia, allowing you to live your life one inner self through another, making possible your synchronous reflexive, intuitive, and intelligent cognition. While your soul is also more complex, since souls can have souls who have souls throughout our entire cluster of realities, complicating everything. The current science ignores these, always studying your conditional reflexes and your father's cigars in order to explain how people think, while millions of testimonies describe the higher worlds firsthand in a consistent manner. While if you can project or lucid dream, you can go up there as your soul to see them yourself. Otherwise, you have to wait until you die to learn the truth.

    Currently, if you become a telepath, you will probably enjoy for some time stealing thoughts from the regular people around in order to learn their secrets, since it might be fun, or you will learn to communicate with the other telepaths over long distances in order to avoid phone bills. However, as a telepath, you do not have to communicate only in speech, through normal sentences, as you do here in this world, but you can interconnect in entire created worlds where you experience everything comprehensively, not only speech. If you want to tell the other telepaths of entire occurrences, stories, and circumstances, you use some words and concepts if you want, yet you can emerge everybody in the entire live story world exactly as you recall it, for a comprehensive experience, with vivid live image, surround sound, and comprehensive needs and feelings, while everybody helps you define all details of your co-created world in every manner they know and understand everything, making the entire story very vivid. While you adjust this co-created world in even more details, in order to make it more accurate, and this is how you communicate with the other telepaths. As a reference, this is what you do in most of your dreams, mostly if the colors and feelings seem more vivid than in the real world. However, if your common dream world is not of the third intelligent level, but only intuitive or reflexive, expect everybody involved to follow their own stories while dreaming together, lacking consistency. While if you still wonder what all food additives, drugs, and medication do to your cognition, they take away your other worlds and realities, while they kill your other selves

The Human Origins

in mass.

At a closer study, you notice how direct telepathy is normal cognitive human interconnectivity as it involves casually all intelligences of your cognitive system in large numbers. This is how your intelligences live in your inner mind worlds by the zillions, and this is how they interconnect throughout their normal life as they fulfill all their tasks within cellular components, cells, tissue, organs, and bodily systems by the zillions, while sending each other needs and feelings in even larger numbers, forming together the multitude of larger systems of intelligences that expand with ease out in society involving everybody necessary, exactly as needed, because this is normal life. More precisely, comprehensive interconnectivity is normal intelligent human life, and should always be possible.

While you have worlds within worlds within worlds everywhere within and throughout the wider world holding and engaging continuously souls, souls' souls, living human beings, conscious intelligences, subconscious intelligences, and all their inner intelligences and inner inner intelligences as far up and down as lifelines, lines of causality, cognitive processes, and lines of reasoning require, forming the Universal Mind, wider world, Life, or the One, which is normal life.

Why should anyone amputate normal human abilities, as telepathy itself? To make tyranny possible. Because normally, only harmony is human, real, and alive, while tyranny and slavery are not human characteristics. Only consensually, tyranny and slavery become human characteristics, by agreement. While as already seen, you must always agree only on everything that is not already the case in life and in the real world to make it possible, while forming the entire consensus in life and in the world in this manner, by agreement.

Even the telepaths here on Earth and in the worlds above tend to live continuously in co-created comprehensive inner realities even when they simply communicate, while creating this entire world in the process as souls, ending up living here continuously as humans, life after life. This is the origin of this particular world, our world, and these are our continuous creators, the souls, with our Creator at the origin of our entire cluster of realities making everything possible, for all souls and human beings. Yet it seems that telepathy itself along with many other abilities are banned and removed even in the higher worlds, constraining the souls to use specific technology to link themselves. Because you cannot monetize cognitive and social abilities, but only technology, while only by monetizing everything you make all higher tyrants possible.

We also notice a continuous disappointment with the current human behavior and achievement, as these remain of a lower level compared to the higher expectations of our Creator. We also notice an outstanding sophistication of this world and of all higher worlds from our cluster of realities, capable to offer a very high development, abilities, meaning, lifestyle, interconnectivity, and behavior to all humans and souls, far above what humans and souls currently have and use, considerably higher than the zero addicted level and the first servitude tyrannical level mostly in use by all humans and souls. Higher even than the basic animal instincts captivating continuously all humans and souls, to the point where our Creator remains determined to end all his higher and lower worlds, just to get away from the worthless, the meaningless, the harmful, and the unfulfilled, and never again. While now you know both the origins and the end of all our worlds and realities.

These are different than the origins of the wider world and therefore of Life herself, because our Creator is neither the Divine nor Life, but our Creator only stands several worlds below the ultimate supreme reality of Life herself, being integral part of Life as you are, along with all souls and all living beings higher or lower. There are many natural realities above our Creator, going up to the ultimate natural reality of Life herself.

You can also reason within co-created telepathic realities, as you reason within your own replica of the world simultaneously, with large numbers of telepaths at your side enhancing considerably your own reasoning, for very fast and very accurate results. Because your reasoning and entire inner life experience can take place in your own inner replica of the world and in the common inner telepathic world. You must also have your own avatar in order to be there. You can certainly have a self or avatar resembling yourself from the outside world, or you can have directly your soul or anything that you choose. With the difference that common worlds tend to distort everything that you know and are in the outside world, making it be what the entire group thinks and believes that is in the outside world.

If you can elaborate this entire knowledge, you notice how you do not need sophisticated co-created worlds to link you telepathically with everybody else allowing everybody to become significantly more capable cognitively and socially, but you can form a normal living intelligent human society together that links you by default cognitively because this is already part of the human abilities, exactly how you interconnect harmoniously cognitively, socially, and spiritually at home in the family. This is your classconscious intelligence, with your living intelligent social self always present, allowing you a harmonious comprehensive cognitive, social, intelligent, and spiritual interconnectivity with everybody else, all human beings and their souls, which should already be part of a third level intelligent human environment but it is not. While as you study the current consensual society closely, you notice how it works hard to erase the entire third level intelligent human environment everywhere, with the third level intelligent human society included.

If you elaborate this knowledge even more, you can distinguish between entire clusters of created realities and the entire natural living wider world. Because once you link your minds in very large numbers for all reasons outside Life, Intelligence, and the wider world, to tell stories, play games, follow vices, or serve consensually, exactly as it is the case with our entire cluster of created realities and exactly as it is the case with the entire Consensual Matrix, then you fall outside Life and the real wider world. However, once you manage to link all minds naturally while forming normal natural intelligent human societies allowing entire classconscious intelligences with all humans reasoning intelligently together while fulfilling all intelligent human needs and meanings, you are naturally part of Life, Intelligence, Interconnectivity, and the wider world, because this is normal intelligent life in the wider world, always avoiding the Consensual Matrix. While as you study our entire cluster of created realities, it is not an actual wider classconscious intelligence, but only a direct interconnectivity similar to social media, used for entertainment, vices, addictions, tyranny, and servitude. The current Internet is similar, since it does not link computers freely, but it links them only with major networks allowing entertainment, vices, divertissement, indoctrination, tyrannical harassment, and work.

Otherwise, you had more as a human being, since you had an entire third level intelligent human classconscious intelligence spanning the human society, alongside your current third level intelligent family at home, alongside your third level intelligent human conscious intelligence, and alongside your subconscious intelligence that you know well. Therefore, you have the well-known vices, addictions, drugs, tyranny, servitude, and indoctrination that you know well, at all possible subhuman levels. This is why our Creator remains disappointed with humans, souls, and his entire cluster of created realities, eager to end them.

We also notice the entire intention during creation, with our creator seeking to form an entire cluster of realities that could be used by humans and souls in a natural, living, harmonious, intelligent, spiritual manner, yet it ended up used by the Consensual Matrix at all subhuman levels, as you know well. It is similar to you at home, because once you open your front door to all homeless people to come live with you because there is always some free space on the couch, you end up harassed or even dead, since this is always the case in an decayed unhuman world.

Once we take our lines of reasoning further, we notice how

all systems of intelligences within your subconscious mind, conscious mind, and outside in the entire society, cooperate continuously while fulfilling all cognitive and social specialized tasks through a continuous interconnectivity made possible in three different manners or more. First, all intelligences can interconnect through all needs and feelings that they send each other. Secondly, they interconnect by embodying each other if necessary, forming larger lifelines containing several selves, very similar to yours. Thirdly, all intelligences can interconnect by forming together entire classconscious intelligences where they think and act together in an entire common world in a more capable and more successful manner than through simple specialized interaction or embodiment as in the first two cases. Yet it is more likely that all intelligences form entire systems of intelligences through various combinations of these three cases, because everything is possible in Life, Intelligence, and the wider world.

All systems of intelligences can form their own inner co-created classconscious realities where they think together in order to become more capable to fulfill their common specialized tasks within the organism and in the outside world. These are your overall primal subconscious intelligences that you know well, as your primal subconscious eating intelligence, reproductive intelligence, and recovery intelligence spanning the organism throughout their specialized fulfillment as they feed, recover, and reproduce the entire organism, always formed and made possible as overall classconscious intelligences by zillions of subcellular intelligences.

It is similar within your conscious mind, because all your inner conscious intelligences, as the multitude of your concepts and conceptions consisting all your memories and understandings of the outside world, can think, live, and fulfill everything according to their own specialized meaning within your conscious mind. While your current third level intelligent conception of the human origins does the same, offering you all the necessary answers throughout all your intelligent reasoning if necessary, while consolidating itself further throughout the rest of this book. However, it is more likely that right now while reading this book, you are more interested in its structure and characteristics than in its specialization.

More precisely, right now as you read this book, you are more interested in how the human cognition functions, through all these specialized systems of intelligences that can interconnect to live life and reason intelligently with you the conscious intelligence, alongside you, or apart from you altogether, as they form their larger systems of intelligences together while forming their own classconscious intelligences together where they conceive, elaborate, and consolidate their own third level intelligent conceptions at will, even apart from you the conscious intelligence.

However, as a conscious intelligence, you can access and interconnect directly with all your memories in order to reason together, or you can embody them altogether to become them and to reason as them, making them your avatars exactly as the souls make the human beings their avatars by making their conscious intelligences their avatars, while furthermore, if necessary, they make all these concepts and conceptions their final avatars, helping them reason intelligently on all the necessary topics throughout this book.

The word 'avatar' is thousands of years old and it refers to actual people living here normally in this world as avatars of higher beings, many times very important higher beings from our higher realities. This is the normal Mahabharata, but right here and right now in the real world. It happened then and it still happens today, since spirituality, very old history, along with millions of testimonies still point to this specific model of our world that we trace in this book.

Yet this world changes gradually, since the type of souls coming here changes substantially, as they drop in development steadily, while all standards and conditions of this world drop in level accordingly, making this world lower in demands and expectations, and therefore more accessible to everybody yet more unhuman. While you can certainly tell, since the quality of life, lifestyle, and experience decay to the same unhuman levels.

Do you notice how everything about yourself and the world is never happening only on your behalf, but on the behalf of your higher realities just as well? Yet this is only another one of your spheres of interconnectivity manifesting here, your higher sphere of interconnectivity, and it is similar to your own inner and outer spheres of interconnectivity. Now you only have one more sphere of interconnectivity to cooperate within and maintain the harmony, which is your higher sphere of interconnectivity.

As seen previously, the knowledge from this book relates and applies to all your worlds and realities, below and above our Creator. Furthermore, everything that you do here manifests up there as a need, and they fulfill it, since it is received there, and then it materializes here, as part of the correspondent replica of our higher world. Yet everything is done on their behalf up there, for them, and not for us. Because we only help them reason, while they only give us the information and the stage to act on as cognitive intelligent actors, which is this world. This is how we act normally and we live our lives, while whatever happens with us here, whatever we understand and achieve here objectively, it arrives up there subjectively, as a thought, idea, or solution. Since this is their own higher mental model, while for us it is normal living, more or less dreadful, depending on what problem they have up there, and on what they have to fulfill. This is how they use it there, if they find it successful. While they let us fail dozens of times here to persevere before they encounter a successful solution, they only take it and use it, and they get to succeed throughout the higher world, without being harmed.

If it is correspondingly similar to them, they live their entire higher lives on behalf of those above, with those living their higher lives on behalf of those above all the way to Life and the One, then we end up living our lives to serve the Deity, through our casual, normal lives, meaning, and fulfillment. Make it a good one.

People still have successful lives, relatively, if it was not for all drugs, entertainment, ignorance, discrimination, servitude, underdevelopment, beliefs, ideologies, exploitation, and eradication. This is the kind of lower level thoughts and unsuccessful ideas that they get up there, they might find it too loud and they take a pill or they have a drink, and this is how entire worlds and realities die down here including this one, just as you kill your own inner worlds and realities throughout similar addictions. This is how we manage to identify the human end, and not only the human origins, including the origins and end of the human civilization.

5 THE ORIGINS OF THE HUMAN CIVILIZATION

We are switching perspectives now to this world, in order to find the origins of the human civilization here on Earth.

We study first how the specific origins of the humankind still affect you and your genetic line today, shaping the entire society in the distinct pyramid of power that you know well. To do so, we can start everywhere on Earth, since very old records of civilized life are found everywhere on Earth. Therefore, we can simply choose any place on Earth, any place interesting you the most.

We can go several thousand years in the past to study the most significant, the most famous relics mentioned by history, the pyramids of Egypt. Yet everything related to those pyramids and to that specific ancient civilization is fake, deliberately altered by the current science, for various reasons. We will come back to Egypt, Europe, and to the entire Mediterranean Basin shortly, but if we want to study very, very old relics, we have to go to Peru.

Peru seems to offer us abundant old records, which are also ignored or interpreted erroneously by the current science, deliberately. What happened in that specific area of South

America is that the crust and the entire continent sank and remained submerged for a long time, even for millions of years, but then it rose and it came back to the surface a long time ago, bringing back above water in this manner all ruins and relics of those very old times and civilizations, for us to study and learn more about our very distant civilization, possibly the actual origin of the Human Civilization.

More precisely, these ruins went under water for a long time, they reemerged with each ice age, went back again underwater with each warm age, while very slowly, the entre crust or tectonic plate rose, and this is how those specific ruins remain above waters today even during this warm age, and therefore you have the chance to study them closely, since their own age marks the age of the entire human civilization.

How old is the Peru civilization? About one thousand years old according to history, or several thousand years older according to alternative researchers ignored by the current science. Yet it is significantly older, tens or hundreds of millions of years old, or more. Similarly, the pyramids Sphinx of Egypt are also older, since civilizations have been around for a very long time, while only the most recent civilizations are human.

These ruins and relics must be far older, but how exactly does everything relate with water levels? What were our very old civilizations doing underwater? Do humans actually come from fish, and not from chickens and monkeys as it is currently believed? The current science associates humans with their most outstanding achievement, standing on two legs. Yet since only humans and chickens can do so, now this must be the combined human and chicken evolution, according to the current science.

In Peru, you can find ruins and artifacts millions of years old and older, right there, laying in the dirt everywhere, and nobody studies them. While if you find that in Europe, you make the evening news. Because civilization in Europe is officially considered to be older than everything else, because Europe is prettier, while in this manner, you are allowed to find older ruins and older artifacts in Europe. Similarly, if you dive not too far off the coast of every continent, you find Atlantis, genuinely. You find it off the east coast of Americas, or out of the west coast of Africa, west coast of Europe, in Oceania, everywhere in the Mediterranean Sea, and in many other places on Earth now submerged. Very old structures and artifacts are everywhere, with most of them hidden well by the current consensual society, in order to hide your past.

Atlantis might not be the name of one city or one entire nation from the past, but the name of an entire age of Earth, Atl-antis. 'Atl' means world, similar to Atlas. While 'Antis' means long before. Atlantis means the world before this world or the age before this age, literally, while antiquity means the same, the world before they died, or the age before this age. However, since the first records of Atlantis are mentioned in antiquity, Atlantis is the antiquity of our antiquity, or the antiquity before the antiquity of our antiquity, placing Atlantis several worlds before this one. Yet since there are only five or six ages in total, some golden and some dark and dreadful, Atlantis could not be too far in the past, only a few ages before this one. We are interested in all these previous worlds, ages, and civilizations, while we are also interested in their origins, and in the origin of the first age or civilization. Whatever the case, notice how the age of our combined human civilization is hundreds of thousands old, or millions of years old.

Atlantis comes from 'Atlantida,' which means antiquity, since it is the same word, only used in different languages and in different civilizations, in the old Latin and in the old Greek, both referring to the same thing, the world or the age before, yet you never know how many ages in the past, since all the old records are erased systematically in the current consensual society.

You have to know everything, while the more you search for the human origins, the more you find in the past additional origins. As the origin of this age or world, the origin of the world before and of all worlds before this one, for all regions of the world, for humans and for life in general, or for the world in general, for the universe, for all realities, and for all intelligences inner and higher.

You already know of Lemuria or Mu, the world before Atlantis. Through its name and beliefs, Lemuria points back to another old age, the Stone Age, yet it seems that Lemuria was the Stone Age altogether. Additionally, humans and humus are words derived from Mu, which means Earth in the old language spoken on Earth long before Atlantis, which is several Earth ages before this one, if you can still keep track. Furthermore, hu-mus, and hu-man, mean something and someone belonging to Earth directly, inhabitant of Earth.

These two distinct ages and the two distinct civilizations that they formed: Atlantis and Lemuria, seem to replace and succeed each other, showing up in various places of Earth and in various times, while we must know why.

Yes, why having one age, then another, then another, then another, then another, and then us, the 'modern' age? Why not having a continuous, prosperous, unique human civilization? Because the environment of Earth should always allow life, societies, and civilizations to develop in the most original manner continuously, but not only successively. Yet what we witness in the human history is a persistent succession of birth - death cycles, since this is the actual Phoenix bird, it is our world, the human world.

What causes the beginning and the end of all human ages, worlds, and civilizations? Everything, including Higher Laws, the natural law of action-reaction, local and global conspiracies, fulfillment of only lower needs as servitude, tyranny, and addictions instead of everything intelligently human, fluctuations in the temperature of the Sun, global warming and global cooling, fluctuations in all astronomical values involving Earth, significant global fluctuations of water levels, volcanic eruptions, falling asteroids, fluctuations in the Earth's albedo distribution, more global warming and global cooling, global warming and global cooling again, and then more conspiracies, more addictions and dynastic tyranny and servitude, then more Higher Laws, and so we reach the 'modern' age, with

The Human Origins

everything mentioned above still happening undisturbed, and how surprising. With humanity caught in all these while struggling to make it through, with the entire life on Earth struggling similarly, civilized and not, and probably along with the rest of life in the Solar System and further away struggling similarly, civilized or not, since this had always been the case since ever, repeatedly, and even cyclically, while nobody cared then, and nobody cares today. Why bothering with all these in a continuously decayed world? While only Atlantis seems to have been different, the famous golden human age, when everybody developed substantially, even past the third intelligent human level.

If the human civilization is older than what science states, why don't we find relics and vestiges everywhere we live? Why having only the Pyramids, the Stonehenge, and the Baalbek in the world to show us the past? Because as we already notice, humanity never lives in one place or another on Earth to settle there permanently and make life easier for everybody, but humanity moves around continuously throughout the ages, literally, while chasing its own tight human niche, as this seems to move around the world, making everything more complicated for everybody, for them in the past while moving around, and for now currently as we trace their steps.

What happens is that, throughout the ages, literally, humans are forced to move around the planet, in order to avoid sometimes rising water levels, and sometimes ice. While time, nature, the elements, science, and the rest of conspiracies destroy all the old human traces on the way, making history very simple and very easy to learn in school: nothing happened here this entire time besides our current age, and therefore there is nothing to see here in this entire section of the museum, so move along, move along. There was nothing here on Earth during the last millennium and any time before, so there is nothing to study today in school, again.

More precisely, during some ages, ice covers Europe, Asia and North America above the 45 degrees latitude, more or less, leaving only Africa, Australia and South America uncovered, while uncovering even Antarctica at times, but only during some ice ages. Then during other ages, it is the other way around, since ice covers the Southern Hemisphere, uncovering the north, as Europe, Asia and North America, making life possible there, which is also the case in this current warm age.

Since this is how the human civilizations form in one place of Earth but then humans have to move and live elsewhere because of all climate changes, while forming other civilizations. While all climatic changes are cyclical on Earth, they depend mostly on Earth characteristics, and this is how the human civilization moves around, similar to Earth itself.

How often does this happen? There is a 26 000 years cycle generated by the precession of Earth, but there is more to consider, from the change of the elliptical shape of Earth's orbit to the temperature of the Sun that seems to fluctuate significantly both cyclically and continuously in time. Ice ages do not only cover most of the continents in ice, making life impossible there while destroying traces of older civilizations, but since the water of oceans gathers around the poles through precipitations and so it freezes there, ice ages lower the sea levels significantly, by hundreds of meters at times, or by about two hundred meters average more precisely. Seashores become fertile grounds and they offer the main niche for humanity, and therefore civilizations develop mainly there, on the lower seashores, and prosper. Yet at the end of the ice age, when the Earth warms up, the ice melts everywhere, the sea levels increase again, slowly, gradually, hundreds of meters, century after century, forcing people to expand cities and settlements on higher lands, away from the rising water, spreading out and covering the freshly available ice free continents. While as you notice, ice and water manage to cover repeatedly exactly the places where humans lived throughout the previous age of Earth, erasing all traces of human past civilization. While humans cannot change the climate of Earth, regardless of how much they try, and therefore they move around from one age or civilization to another.

More precisely, what was first covered by ice during ice

ages, is now available during warm ages, ten thousand years later. While the lower lands once habitable during the ice age around the lower coasts of all continents are now unavailable, all being hundreds of meters underwater. Yet even if water cannot reach to flood these lower lands, now they are hot deserts voided of life.

More precisely, people live in specific parts of Earth during some ages, and then they live in entirely different parts of Earth during other ages, but not in both, while they always do so as distinct civilizations. Even so, humanity should have been able to assure a smooth continuation of the human civilization, even though these areas are either covered by ice, deserts, or water, depending on age, yet humans are not able to continue their own civilizations, always for the same reason, lack of proper human development.

Because surprisingly, you cannot take drugs continuously while also avoiding death and extinction since it is impossible, and you always die. Similarly, you cannot rule tyrannically from one dynasty to another while still expecting humanity to prosper and avoid extinction, since again, it is impossible. These two types of ages succeed endlessly, the warm and cold ages, sometimes lasting for thousands of years, while other times lasting for tens of thousands, hundreds of thousands, and even for millions of years at a time, just study the records to see it for yourself.

Did humans pollute the Earth endlessly, to have caused the succession of all these ages through greenhouse effects and global warming? Certainly, or this is what the current science could also state, but where are all the old tires? There is more to consider than car tires, since the temperature of the Sun fluctuated cyclically while causing most of these longer-lasting ages. While the Sun cools down very slowly throughout its lifetime as all stars do, so expect longer-lasting ice ages in the following few billion years, because five or seven billion years from now, who knows, it could even make Earth similar to Mars.

Atlantis was the name of the civilization that flourished

during the last ice age at lower sea level, found underwater now. Lemuria was the civilization preceding Atlantis, which flourished here where we live now, before the ice of the last ice age covered the continents. Part of Lemuria, along with all Giants and all the mammoths died then, with that everlasting snowfall that started suddenly twenty-two thousand years ago. This explains why we do not find very old vestiges and structures wherever we live.

At the end of an ice age, water levels rise, covering low lands very slowly, four meters per century, and therefore giving enough time to everyone to move away and not sink. People do not even have to walk away from the rising waters, since they cannot even notice the rising waters, taking place very slowly, throughout thousands of years. People only build their new dwellings on higher, available land, while remaining ignorant of the entire transition.

It is said that during the last deluge, Atlantis sunk in only one night. Some people decided to die there, while others left. Currently, there are some nations in the world, people without land, who could have migrated away from these low lands. This was the case if the rise of water was sudden, taking only days, weeks, months or years, causing these people to lose their land and belongings within one lifetime.

Even their names associate to these lands. 'Giza,' 'Gaza,' 'Egypt,' and 'Gipsy' have a common part, a common sound: 'Gi,' 'Ge,' or 'Ga,' as 'Giant,' 'Geo,' 'Geea,' and 'Gaya,'. 'Gaya,' 'Geea,' and 'Geo' are always the old name of our planet given in these specific places, while 'Giant' means literally inhabitant of Earth, from the suffix 'ant,' which means 'inhabitant of' or 'belonging to.' 'Giza' and 'Gaza' mean high land, with 'za,' 'sa,' and 'su' always meaning high, top of, illuminated by, or enlightened.

'Egypt' defines a country today, while 'Gipsy' defines only people, nomadic people for a long time, but where is their country? That country was called Gypt, Gia, or Geea back then. It is underwater today, and it is the Lower Egypt. The part 'psy' from 'Gypsy' relates to thoughts, minds, and higher cognitive abilities, similar to **psy**chology and **psy**chic. Gypsies are still psychic today, or some still are, because medicine and conspiracies in general take the higher abilities from them continuously, the way they take away your own higher abilities, if you were supposed to have any, and mostly if you have a higher self within your cognitive system to represent your higher powers.

People in the Lower Egypt believed in and venerated the Snake Goddess. The venom from snakebites was used in the Egypt of ancient times as a hallucinogen, the way specific mushrooms, DMT, salvia, Ayahuasca, and LSD are used today for the same purpose, among others. Other nations used and still use specific plants, berries, vines, leafs, slime from the skin of specific frogs, strong meditation, and intense rhythms and dancing. These helped ancient people engaging their higher abilities while transcending and projecting, while also using the venin of specific snakes. Cleopatra killed herself and died bitten by three of these snakes, in a specific order. Study these snakes along with the entire ritual associated to them, to find that Cleopatra might have not died, but she transcended systematically to a specific place on the firmament or in a higher reality, the place where all pharaohs went.

All pharaohs had always claimed that they belong to another place, and that was their own origin, or your own origin, an objective or highjective place somewhere in the Duat region, somewhere in the Orion Constellation or in the higher aspect of the Orion Constellation. Therefore, their souls had to return there after death, and it was very important that it happened. Because they had been imprisoned here on Earth continuously, always seeking to go back home.

Throughout life, Cleopatra made a very strong effort to become a pharaoh, and now her soul had to go or return to Duat. It probably did, just by enhancing her higher cognitive abilities at her most critical moment, when she died. She had also burned the Library of Alexandria right before she died, or this is what science claims today, so the invaders could never learn all higher knowledge of Egypt, all higher knowledge inherited from the Lower Egypt, from the hearth of Atlantis.

There are rumors about some or many people from past civilizations, how they actually came here to escape the Consensual Matrix and prosper. Yet it did not last long, because they became 'greedy,' as the old records state. Consequently, they became discriminatory and exploitive, forming social islands within social islands similar to all social hierarchies and social classes, breaking the comprehensive social harmony, while decaying everyone from their third intelligent human level to the lower animal, servitude, dogmatic, and vicious levels. While these can always take down any advanced civilization very shortly, just look around to see it for yourself. More importantly, this takes place cyclically, through a similar social greed and social decay.

This is how Atlantis ended, perished, or sank, and now you can make the correlation between all events. While you always have people engaging higher abilities on their own only when they can avoid the Consensual Matrix. They learn the truth in this manner, through higher knowledge, this is referred throughout old records as biting from the fruit of knowledge or from the fruit of life. Because the Consensual Matrix can always return, and this is always kept in the human knowledge, as a reminder not to repeat all old mistakes. While the current Brotherhood certainly remembers it and treasures it accordingly, also repeating all old mistakes, but always for a good cause, which will be revealed soon, while ruining this entire world. Yet since this is the last world, it is terminal this time, no more.

The Gipsy people of Europe are relatively psychic, while they still make money foretelling the future and the past, many times by reading Tarot Cards, which have also originated in those old parts and old times of the world, along with crystal gazing and the touch of people and objects in order to foresee and foretell events. These are the Gyptians of the distant past, not the Egyptians, because according to the old records, Egypt and the Lower Egypt were two distinct nations, and did not resemble. Similarly, there is a distinction between the people of the upper Italy and lower Italy. While it is the same in Scotland and Ireland, as they always distinguish between high lands and low lands.

You can certainly make an idea by now of what genetic lines have higher selves and higher abilities, and what genetic lines do not. The question is why this is the case, since all humans are humans, and should all have similar abilities, both cognitive and physic, while you already know the truth. Even more, most of the medication, vaccines, and food and drink additives are meant to kill your higher abilities, which are actually specific higher intelligences part of your cognitive system performing these higher tasks. Science, medicine, agriculture, and food industry undergo a big effort to distribute these substances to each individual daily, to the young and the old alike, and this is what the human civilization had become. How can humans ever survive without all their abilities? They cannot, since they never could from one age to another, yet now it is too late anyway.

Civilizations occupy specific niches, and consequently, they need specific niche elements in order to survive and develop, as food, water, metal, stone, wood, bodies of water, and clear land. Deltas offer a very rich land for agriculture, generating abundant food, enough for a large population. The Nile Delta was the best place to live during the last age, while among all places, the Ancient Lower Egypt must have been the strongest and the most developed. Did you know that the Pyramids and the Sphinx delimited the Lower Egypt from Egypt? This means that Giza, or part of Giza, including the sphinx and pyramids, belonged to the Lower Egypt, to Atlantis. The Sphinx had been remodeled many times throughout history while deities changed names, shapes, ideologies, and appearances, yet originally, from the length of the body and the skin around the head, the Sphinx represented a snake, the Snake Goddess, the main deity of the Lower Egypt. The Sphinx still resembles a cobra from the front. It has a different, remodeled head today, yet it is small, the body is long, it is a snake, or even a snake with legs, which is the Snake Goddess.

The ancestors of the Gipsy people of Europe might have inhabited all the lands around the Mediterranean Sea, now underwater: the lands north of Egypt, the lands south of Turkey, the submerged lands of Greece, along with the lands south of Europe now covered by water, since they resemble closely the people of Egypt, while resembling slightly the ancient people of Canaan.

People's resemblance is a very important characteristic when studying people's origin. Canaan was wider then. Gaza strip reached one hundred kilometers offshore in the past, same as Israel, spreading all the way west to the sunken parts of the Lower Egypt, the sunken part of the Nile Delta, sometimes spreading farther away, sometimes closer, depending on the ice age. I do not claim that anyone has origins in sunken lands, since people moved, travelled, migrated, married, emigrated and immigrated everywhere and from everywhere in the area of the Old Egypt and Old Canaan. It was even easier for people to migrate then than it is for people to emigrate today. People also traded slaves everywhere, while even free people sold members of their families and themselves into slavery, everywhere they wanted. Even more, people inhabiting these lower lands of the Mediterranean Basin were renowned sailors and mastered the Mediterranean Sea, taking everybody and everything from anywhere to everywhere. This is why today, genetic lines and people in general have diverse, wider origins. What does Mediterranean mean?

It is sad how we have only words left today to help us understand facts and events of the past, with nothing else to uncover and sustain the truth. I am studying here powerful ancient civilizations, very rich in both wealth and culture, and there is nothing officially accepted today to confirm their existence. Where are the ruins and relics? Who hides them? What is left today in the open is still there because they were too important religiously, as the Bibles, or they were too big to be hidden or erased, as the Pyramids. You are aware of the very old tunnels and chambers underneath and underground found everywhere near the Sphinx and Pyramids in Egypt. An entire ancient library was destroyed in Alexandria only two thousand years ago, all burned down by the Egyptians themselves it is said, or by Cleopatra herself before she died. Valuable, very ancient knowledge is said to have been stored there, all gathered from that area by Alexander the Great. Alexandria is the closest city to the sunken Nile Delta. That area was exactly what Plato referred to as Atlantis, the powerful civilization, being exactly to the north as he stated, very close to where he lived. Divers still find impressive artifacts in that area, all resembling in style the art and architecture of Ancient Greece and Ancient Rome. It is said that there is an entire sunken city right north of Alexandria, an impressive necropolis. If you go east from there, a few hundred kilometers, you reach Jerusalem, which is also said to be honeycombed with ancient tunnels. People also found very old artifacts rich in art and knowledge left from very old civilizations, before Mesopotamia.

Mesopotamia is currently the officially accepted origin of our civilization, the very first civilization of Earth, yet as you study closely all vestiges found in and around the Mediterranean Basin, you find them older than Mesopotamia itself. North of Jerusalem, at Baalbek, the ancient buildings are impressive. Their architecture is unlike anything we have today, with gigantic pillars of stone impossible to be cut and transported by humans, while being used to make simple walls. South of Sicily, the seafloor is high, and it was out of waters during ice ages, spreading all the way to Africa. A powerful civilization also existed there, and now Malta is what is left, an island. It is said that Malta is also honeycombed in very old tunnels, with some leading all the way to Europe, if they are not flooded.

Why tunnels everywhere? Because humanity dies sometimes by draught and sometimes by ice as it happens throughout warm ages and ice ages, yet there are other cyclical dreadful events ending the human civilization, as intense solar flares and intense ultraviolet radiation coming from the Sun right through an unprotected atmosphere, which is always the case during the reversing poles of Earth and through a multitude of other occurrences, while you have to move underground whenever these happen in order to survive. Even so, if only a small part of the human civilization survives, as only the elite, they still perish, since the elite is always worthless. The elite is incapable to tend to itself, and it can never survive on its own.

The Consensual Matrix can inform you in advance of all incoming natural and artificial disasters as it always does, but only if you are faithful enough and if you serve well. While you can know the future in advance yourself, through your higher abilities, if you manage to live your life on higher developmental levels and outside the Consensual Matrix. While in this manner, only the golden ages of the human civilization survive, while the rest perish. Because as stated, you cannot rule the world tyrannically as the current hierarchic Brotherhood and the Elite do, while still assuming that you help the world, because you always perish with the first major natural disaster, as it is the case with all tyrants, dictators, megalomaniacs, and entire hierarchic Brotherhoods, dynasties, and invisible kingdoms. Die away.

One can only imagine the amount of ancient information left from this extraordinary civilization of the last ice age, yet everything we have today in the open is one single reference to Atlantis made by Plato, randomly, in one of his books. While even that statement by Plato might have been accidentally left intact, because might have written entire books on Atlantis, now destroyed deliberately to hide the past. While the current consensual science never studies even that slight statement about Atlantis, while accepting only Mesopotamia as the origins of the human civilization, Mesopotamia itself, since it was already in the Consensual Matrix, with the rest of the world free and significantly more developed, mostly in and around the Mediterranean Basin.

We also have gigantic structures left from those very old and very developed times, both on land and underwater, and

The Human Origins

nobody can explain what they are, who made them, how they were made, and what they are doing there. It is said that, before these civilizations sunk, people constructed specific storing chambers everywhere on higher land, and filled them with records and artifacts, saving them from flooding, for their future descendants to learn who they were and where they came from. While all their descendants work hard to erase and destroy who they were and where they came from, while serving diligently.

Where is everything now? Because I searched everywhere for anything accurate enough to state here as an officially accepted reference to all previous ages of humanity, and I found nothing. Which means that Atlantis never existed, the way science already states, and therefore there is nothing to learn in your history lesson at school. Just study the World Wars all over again, along with the Civil War as you always do, and you graduate.

How rich were those lands in the Mediterranean Basin, now covered by waters? They were rich enough to generate the beautiful style in art, writing, and architecture of Ancient Greece, Phoenicia, Egypt, and Ancient Italy, which are the main spinoffs of Atlantis, very original and unique, unlike anything else in the world. These spinoff civilizations were the major inheritors of the past golden ages of Earth, spreading from there to the entire Europe, and still present today.

Deltas can create a multitude of nested islands, while it is very easy for the inhabitants of any delta to divert smaller and larger waterways in order to form isolated, protected, fortified territories, which are technically islands. This was the case with the entire Nile Delta of the past, currently mostly underwater, and with the entire Tibur-Euphrates Delta that happens to have held the homologue civilization of the past, the Sumerians, along with all their precursor and successor civilizations. The Sumerian civilization was more than a homologue civilization of Atlantis, since it could have been a direct precursor, or even a master or controlling civilization. Yet you cannot know the truth, since all records are hidden or destroyed.

Why exactly hiding the records of an entire past human civilization, highly prosperous from all standards? Because you cannot have prosperity and bondage simultaneously, so choose one. Yet there was not only one prosperous civilization in the past but more, since you always have prosperity, equality, and development every time the Consensual Matrix is absent from Earth. While the Consensual Matrix remained absent throughout many of the past ages of Earth, allowing all humans to develop beyond the current level.

Yet you still have all the past golden ages of Earth in you, because all developed human intelligences are still in you, since primal intelligences never die, but they transfer themselves wholly from one generation to another at the moment of conception. While now they send you exactly the developmental needs to make the world exactly as prosperous and egalitarian as it used to be, you have it within. However, other nations of Earth went only through dynasties continuously, and now as you study them closely, you find them less developed, and always eager to serve. This is why you have dictators mostly in the East, yet they fight hard currently to take the entire world, while decaying everything. No more golden ages, and no more humanity.

These ancient civilizations were powerful enough to keep their integrity throughout millennia. The Lower Egypt had remained an independent nation, having its own language, writing, culture, tradition, beliefs, religion, and art, independent from the Upper Egypt, which is the current Egypt. E-gypt means Higher Gypt, or the current Egypt. Yet the two countries united five thousand years ago. Afterwards, the people from the Lower Egypt still kept their language, traditions, religion, and culture intact from the highlanders, the current Egyptians. Why exactly did the two nations unite so suddenly?

When sea levels rose, suddenly or not, during one of the most famous historic events, the sinking of Atlantis, if the fleeing people of Lower Egypt went south to the Upper Egypt, the Egypt that we know today, they might have become slaves. This is consistent with the Bibles and with all historical records. Yet the people of this lower Mediterranean region must have migrated everywhere, even to Asia, for the close resemblance between the people of India, the Gipsy, and the people of today's Egypt. Do a fast research, to see that the ancient Indians looked different from today's Indians, since they were taller and larger, with Asiatic features, as a lighter, yellowish skin, larger heads, and larger oval eyes, all being genuine Asiatic details found naturally in that specific place of the world, integrating naturally there. Where are these people today? Ancient Higher Egyptians looked different, they were also larger, with rounder faces, larger heads, larger rounder eyes, darker skin, and they had more African features. Where are these today?

Currently, the invisible kingdom, which are the native people of the Caucasian region, claim their origin in the Canaan region, along with the Palestinians and everybody else living in that area. Yet the natives of the Caucasian region look neither Egyptian, not ancient Canaanite, nor like the Gipsy people of today, but they look strikingly European, or Eastern-European more precisely, having East-European origins. While as you study Atlantis closely, you notice how it expanded eastwardly as far at the Caucasian region, with Georgia included, marking the outskirts of East Europe, while placing the origins of the invisible kingdom at the outskirts of Atlantis, but not at its hearth. While the old Canaan was more towards the center of Atlantis, alongside the Ancient Italy, Lower Egypt, Phoenicia, and the Ancient Greece.

Currently, the entire Caucasian region seeks to join Europe and therefore to become genuinely East Europeans. Yet as stated, the entire Caucasian region was already part of that very prosperous Mediterranean civilization called Atlantis, right at its very east, in the old East Europe. The Caucasian region was part of Atlantis during the last age of Earth, prospering within the same very advanced civilization of Earth, alongside the current Italy, Egypt, Lower Egypt, Canaan, Alexandria,

Baalbek, Gibraltar, Southern Spain, Greece, old Turkey, and Albania, going as far as the eastern and southern basin of the Black Sea, all the way up to the old Istria and Pontus Euxinus from the current Romania.

People are people everywhere, so why is this significant? It is significant for them, for those controlling the world. Because it is very helpful to compete in the current society as an entire social group, since you can succeed better than the individuals living and competing alone in society. This makes all social organizations of the current consensual Brotherhood very capable and very successful, including all lodges, cartels, groups, mobs, clubs, and societies. However, having an entire kingdom or nation competing in society today counting in hundreds of millions, interconnected through a very strong ideology meant to help themselves while discriminating, exploiting, controlling, sabotaging, and exterminating the rest of the world, and while maximizing profit and success at all costs only for themselves, now the rest of the world has no chance to succeed in any manner against the invisible kingdom and fails, including the entire Brotherhood. This is how the Brotherhood failed, only a couple of centuries ago, the old, genuine, egalitarian Brotherhood, being infiltrated by the invisible kingdom, and the world has never been the same ever since. The old genuine Brotherhood became the current hierarchic Brotherhood, right then, two centuries ago, just another consensual tool in the hands of the invisible kingdom, along with finance, science, education, entertainment, industry, history, justice, medicine, insurance, transportation, military, police, war industry, pharmacy, culture, and art, as consensual tools in the hands of the invisible kingdom, along with you and your entire family, whoever you are and wherever you are positioned in the current consensual human society, in the Masses, Brotherhood, or in the Elite. However, currently, all dictators of the world unite against the invisible kingdom, taking it down very rapidly. Therefore, expect everlasting global dynasties from now on, since this is what they plan.

It is important to consider all social actors controlling the world now and throughout history, because these always erase, hide, alter, and restructure the human knowledge in any manner they find more adequate for them, while altering in this manner all information about the human origins, confusing you throughout your research. Because it is enough to place the human origins exactly in the Caucasian region, in Georgia, which is the native land of the invisible kingdom, and now they own the world through normal legal national genetic entitlement. Yet since they also place the origins of the human religion, of the human culture, and therefore of the human civilization in the old Canaan region, since there is where they also claim to originate, now they take everything for themselves, by law, culture, origins, and civilization. Everything is theirs, everything belongs to the invisible kingdom, the entire Earth, according to them.

This is why all dictators unite throughout the world while taking the invisible down, because as all tyrants of the world, they do not want to be owned by the invisible kingdom, as it had been the case for the last two centuries, but they want the world for themselves. With China leading this entire change in world orders, and therefore expect Chinese dynasties from now on at the top of the world, since this is the plan. However, all tyrants, dictators, and megalomaniacs of the world want the entire world for themselves, and they fight hard to take it for themselves, while you never know who wins in the end.

Yet even so, everybody seems to prefer the invisible kingdom of the West instead of the dictators of the East, emigrating in mass to the West, but not to the East. Which is as choosing between Hitler and Mussolini in the upcoming election, yet it could have been better on Earth having no dictators, no exploitation, no hierarchies, and no rulers, but a genuine, egalitarian human society, matching the old, genuine, egalitarian Atlantis, exactly as our Creator had intended.

Yet it is mentioned historically of the existence of en even older civilization than Mesopotamia. It is stated that people came to inhabit Mesopotamia from the higher lands of Sumer, the Sumerians. Where is Sumer? Up in the mountains as the current science states, since the name 'Sumer' itself signifies higher land and it must be up in the mountains, yet there is no trace of any civilization in any mountain from the region. There was only ice up there on top of those mountains during ice ages, and nothing else. Why did Sumerians call them 'high lands,' the way people called Giza and Gaza high lands in the Mediterranean Basin? Because ancient people knew that all lower lands were habitable only temporarily, as long as the ice age lasted. Atlantians knew of the flood long before they sunk. Higher lands meant lands free of water, habitable throughout ages.

Mesopotamia, which means 'the land between two rivers,' appeared and developed along hundreds of kilometers of land south of where it is today, in the lower lands of the Tibur-Euphrates Delta of those times, during the last ice age, lands now under water. People came there from the higher lands, the Mesopotamia or Iraq of today, over twenty thousand years ago.

In the same manner, there are two Egypts: the Lower Egypt, and the Upper Egypt. There are also two Alexandria cities, one habitable today, and the other one sunken right north of it. There are two Mesopotamias today, the Higher Mesopotamia or today's Iraq, and the Lower Mesopotamia, now sunken under the Pacific waters, the part of Mesopotamia contemporary with Atlantis, Lower Egypt, and Lower Alexandria. Yet back then, it was an entire civilization, still above water during the last ice age, and since it was on high land above water, it was called Sumer.

Do you see how the deities or inhabitants of Sumer are very old, dating back to the past ages before this one? These were in the Consensual Matrix, in contrast with Atlantis, which seems to have remained free of the Consensual Matrix, until right before it sunk. You can see and read it in their tablets, and this tells how old, how powerful, and how persistent the Consensual Matrix is on Earth. While now you know the origins of the Consensual Matrix on Earth.

Yet people do not lose their lands when waters increase in level four meters a century. Could Atlantis have sunk in one night? Could waters have risen very fast, catastrophically in some parts of Atlantis, to leave entire nations landless in only one day? Why sinking so fast? There are three theories, and all could have happened even simultaneously.

In the first theory, Atlantis sunk slowly, due to the weight of its own buildings in the soft delta region of the Lower Nile River, in a Venetian style, since this is how Venice sinks even today, through its own weight, the weight of all rocks and bricks brought in the delta to build the entire city, and this is how it sinks.

In the second theory, the Nile River flooded excessively, catastrophically, at the end of the last ice age, from abundant precipitations contemporary to the famous Deluge from the bibles, and from the fast melting of all glaciers of all mountains of Africa, washing away an entire civilization. The people who left first were the ones to survive, while others stayed behind, from patriotism, discouragement, or greed, until it was too late.

This flood could have taken place in one night, or even faster, yet how could the people have known of the exact occurrence, the way records still state? Where and how did they leave? They had to use boats, since the entire delta was a big deluge for hundreds of kilometers, and you could not walk directly south to the Upper Egypt. Did it also rain? Yes, there is a large amount of rain taking place at the end of any ice age, lasting for hundreds or even thousands of years. What has become of these people? Some settled on higher shores, and Canaan was the closest, with the Gaza Strip and the kingdom of Israel to the west. Boats can take you everywhere, so people could have gone north to the Ancient Greece and South Europe. Some people went west to North Africa, others found their way south and continued inhabiting the higher lands of the remaining Lower Egypt, while most of them could have remained living on boats, making for the busy traffic in the Mediterranean Sea of those times, since all advanced cities and civilizations were everywhere around the Mediterranean Sea at that time, on all shores. These people were called the Sea People, they became stronger gradually, and a few thousand years later, they invaded and conquered both Egypt and the Hittites at once, two of the most powerful nations of the region and of the entire world, while this tells something.

In the third theory, the Gibraltar Strait floor was higher than the level of the Atlantic Ocean during all previous ice ages, and it stopped the Atlantic Ocean waters from flowing into the Mediterranean Sea at the end of each ice age. Therefore, the waters of the Mediterranean Sea remained at a constant very low level even during warm ages, keeping the Mediterranean civilization continuously instated from one age to another, while allowing the people of that entire region the necessary time to form a very strong and a very prosperous civilization, age after age: Atlantis.

Today, the Mediterranean Sea is saltier than the world's oceans, which means that water evaporates faster in the Mediterranean Basin than all oceans and than the Black Sea, faster than what the fresh water from rivers and precipitations can replace. Precipitations are even lower during ice ages, and it is estimated that the entire Mediterranean Sea loses water fast and it can even dry up completely in a matter of centuries, as it is mentioned in the Bibles, with pockets of very salty water remaining in some areas, wherever the basin is deeper.

The most fertile lands are the deltas of major rivers, as the Nile Delta, hundreds of kilometers wide to the north, able to nourish a very large civilization. There are traces of the Nile River on the Mediterranean seafloor 1500 meters deep, yet they could or could not have come from the last ice age, but from any other ice age before the last one. You can see these traces yourself in any satellite picture of the area, as a multitude of sunken riverbeds found at the outskirts of the submerged Nile Delta, now deep underwater. The sunken part of the Nile Delta is up to 2000 meters deep, and it could have been out of water during the last ice age totally, with the presumed natural dam of the Gibraltar Strait still in place. Without this natural dam, the Mediterranean Sea level was only 150 meters lower than it is today, with the Atlantic Ocean communicating freely with the waters of the Mediterranean Sea.

If there had been a natural dam there holding the waters of the Atlantic Ocean from flowing into the Mediterranean Sea for thousands of years throughout the warm ages, then the Mediterranean Sea level was even 700 meters lower than what it is today, maintaining a constant low level throughout all ages. Therefore, any civilization developing in the Mediterranean Basin had enough space, time, and natural conditions to flourish undisturbed from one age to another for a very long time. But one day, when the ocean level became too high, they finally spilled over the Gibraltar natural dam and flooded these lower lands, bringing rapidly the Mediterranean Sea at the same level with the world oceans, hundreds of meters higher, while sinking an entire very old, very strong, and very prosperous civilization.

While even the old mythical and historical records state these, with Hercules himself conquering Atlantis by flooding it entirely in this specific manner, and the Greeks won, while also flooding and killing themselves, since many of the Greek islands also sank right then with the entire Atlantis.

Why would you ever seek to conquer an entire very developed civilization? For honor and virtue, as all ideologies determine you to do? All tyrants define honor and virtue closely, exactly as it is more conveniently for them, and now this is what you believe, while conquering the rest of the world in the process, just for them. Therefore, why exactly would our Creator keep you and your entire undeveloped world around, mostly after he lost his beloved Atlantis? Yes, Atlantis was the land where gods walked with men, so why would our Creator build a newer and better world for you so trash similarly, as you have already done with the six previous ones? More precisely, how many worlds, origins, ages, and human civilizations do you want to have while still taking drugs and while still serving all tyrants? Six? Ten? One hundred? Two hundreds?

This could not have happened in one day, yet when waters flooded and spilled over the natural dam in the Gibraltar Strait, it eroded it very fast, waters rushed in with an unimaginable force, digging away that dam from the top down in only days or weeks, while make a gigantic waterfall and an impressive spectacle, filling up the entire Mediterranean basin in a relatively short time. If this happened six thousand years ago and not further back in the past, then it could have made the once-prosperous inhabitants of the rich, lower lands between Africa and Europe to leave in a hurry, lose their lands to the rapidly rising waters, and become nomads as the Gypsies, part of them travelling from place to place throughout Europe, Asia, India, Africa, and the Middle East, and part of them settling everywhere they were allowed, as you still find them everywhere today. While they are very numerous, of a unique genetic line, and therefore of a unique nation.

This was the case with all people inhabiting the entire lower Mediterranean Basin during ice ages, only that the people inhabiting the Lower Nile Delta had the chance to be significantly more numerous and more advanced and prosperous through the high fertility of their vast land, the Nile Delta, capable to feed a large population, and therefore capable to make possible a very large and very advanced civilization, Gypt as it was called in the distant past, or the Lower Egypt as historians call it today. While judging from the scarce records still around, they had higher human abilities, a great achievement for any civilization.

As a reference, the national symbol for the Lower Egypt was not only the Snake Goddess, signifying advanced intelligence and advanced higher psychic abilities, but also the papyrus flower, signifying writing, research, written records, arts, and science, the actual human intelligence. This is what the Library of Alexandria held, these written records or only what was left of them, as they had been found scattered throughout the entire northern Egypt by the people of Alexander the Great.

Was there a natural dam on the floor of the Gibraltar Strait holding the Atlantic Ocean for ages from flooding the Mediterranean Sea? You can study all maps with ease right now, mostly on the Internet. The Gibraltar Strait is about 16 km wide and 50 km long. Waters increased 120 meters at the end of the last ice age, while the Gibraltar Strait is more than 700 meters deep in the Mediterranean side, going down to one thousand meters deep and then two thousand meters deep very shortly, which is the actual depth of the Mediterranean Sea at its two centers. If the waters increased only 120 meters at the end of the last ice age, while the Gibraltar Strait is over 600 meters deep in the middle, we cannot consider a natural dam there in the past. Yet if it had ever been present there, it had certainly eroded in a matter of days or weeks when the Atlantic Ocean overflowed in the Mediterranean Sea, and there is nothing to see today.

Yet we can certainly tell, since far in the Atlantic, on the western part of the Gibraltar Strait, waters are very shallow. The ocean floor is as deep as only 250 meters there. It does look unnatural for the water exactly in the middle of a strait between two continents to be three times deeper than further offshore in the ocean. Even more, when you study the strait, the way its floor is shaped, you notice that it had been massively excavated, just the way water excavates massively entire canyons through receding massive waterfalls. Since this is how all canyons are formed, by massive waterfalls and by their vertical water erosion. Just study the maps right now to see it for yourself.

There is a massive canyon in the Gibraltar strait right now, underwater, formed by a massive waterfall, by far the largest in the world ever, when the ocean waters managed to spill into the Mediterranean Sea. This happened only once, at the end of one of the major ice ages to have covered the north hemisphere. Ocean waters filled the Mediterranean Basin relatively rapidly then, to submerge Atlantis, this entire civilization that lived and prospered there continuously throughout all the previous ages of Earth, undisturbed by the significant variations in ocean levels that had displaced the entire world repeatedly throughout time.

There is an old legend describing how Hercules had conquered Atlantis by forcing open the mighty gates of the ocean, letting the waters pour in, and sinking Atlantis. Any nation at war could have dug a small ditch in a matter of months, to allow the water of the Atlantic Ocean in the old Mediterranean Sea, and then the entre dam eroded rapidly, 600 meters deep, in a matter of days, weeks, months, or more, as it transformed into a canyon, exactly as you still find it underwater, in the Gibraltar Strait. You can study the maps to see it yourself.

However it happened, there is a difference of 500 meters between the depth of the Gibraltar Strait on the Atlantic side and on the Mediterranean side, which is the perfect shape of an ancient, massive waterfall of 500 meters high and 40 kilometers wide. The massive canyon left behind is still underwater, which not only confirms the entire deluge, but by its shape, it states that water level was over 700 meters lower in the Mediterranean Basin before the last flood, making for a very wide habitable fertile land in the Mediterranean Basin during the last ice age, now underwater, the perfect habitat for a very prosperous civilization, the Atlantis before Plato. Even more, since most of Europe was under ice, the Mediterranean Basin was among the few available fertile lands of those days.

This is hidden from you today, an entire human civilization lasting for millions of years, probably living life at the intelligent human level most of the time. You can always tell when you live life at the intelligent human level, because you develop consistently socially, cognitively, technologically, artistically, and spiritually, ending up with extraordinary higher powers, while these higher powers keep you successful and developed at the intelligent human level for ages in a row. This is how the people knew when ice ages and warm ages succeed each other, while knowing which lands remain habitable, marking them accordingly.

This major disaster might have not happened exactly at the end of the last ice age, but a few thousand years later, when the ocean level became high enough to spill over the dam of the Gibraltar Strait. From all records and from all Bibles and Egyptian records, we learn that this catastrophic event happened about 6000 years ago. Yet since all old records including science are altered to show a very recent past, now you can never know when everything really took place.

Yet you still do, if you are in the invisible kingdom, since the calendar that you use in the invisible kingdom is about 6000 years old. More precisely, the current year is about 6000 in the invisible kingdom, depending on when you read this book, marking exactly the origin of the last age, the Antiquity, or Atlantis.

This forms a basis for ancient civilizations to exist in the lower lands of the Mediterranean Sea, while this proves that civilized life on Earth continued throughout ice ages. Which implies that our current civilization is much older, and we were not animals only thousands of years ago as history states, but normal human beings, living a normal, civilized life. Or we were even more civilized and more developed in all aspects, living life at the intelligent human level and beyond, mostly if the Consensual Matrix was not present on Earth.

This is what science struggles to hide from you, highly prosperous and highly fulfilling ancient times lacking the Consensual Matrix on Earth, with all humans integrating harmoniously through all their spheres of interconnectivity with everyone and everything around, above, and within. Mother Nature. Since this is the actual human lifeline of existence, along with the intelligent human society, the overall human family, and the genuine egalitarian Brotherhood of the old times. While you cannot achieve these today consensually, because the Consensual Matrix is of the first developmental level, while humans are by nature third level intelligent living beings, living in genuine third level intelligent human worlds. While whatever you have today is not a human world but something else, whatever is agreed throughout the Consensual Matrix for you to have here on Earth.

What else do we know today? Officially, the Mediterranean Sea is accepted to have been completely dried six million years ago, according to sediments. Officially, the Mediterranean Sea level did descend and ascend with each ice age, sometimes more and sometimes less. Yet this could have never influenced anyone, since nobody lived in that area anyway, the way the current science states, and everybody was still in the Stone Age, if people existed at all back then, while inhabiting the upper, colder dry lands, the lands that we happen to inhabit today, but never underwater. The current consensual science is modular, stating on one side that the Mediterranean water levels vary 120 meters with each ice age, yet on the other side, while studying people and entire nations in the area, the water levels never vary, and therefore nobody ever lived in the Mediterranean Basin, because they could never go underwater.

The Pyramids and the Sphinx are officially dated by science at four or five thousand years of age, long after the deluge. Yet there are still people studying the water erosion on the Sphinx, stating that the Sphinx has to be much older than the end of the last ice age, since then, at the end of the last ice age, it rained abundantly for the draining water to carve the rock of the Sphinx. Yet the current science ignores them, since it has a very well defined agenda to follow, and this is what it discovers. What agenda exactly?

We are studying now the moment in time when our history had been chopped off, truncated, and hidden away from us, along with our roots and origins. Who can censor history at this massive level? Anyone can do so. Totalitarian regimes can censor everything in their nations, not only history, but all knowledge. This happens currently throughout the dictatorships of the East, as it had been the case not too long ago throughout the dictatorships of Eastern Europe. They concealed and altered knowledge in any manner it was more convenient for the tyrants, and this was exactly what people learned in school and spoke about, with those who said otherwise killed. While as you notice, the more developed people never follow tyrants and dictators and are exterminated genetically shortly, while the undeveloped people are doomed to serve tyrants, dictators, and entire dynasties for many generations or even endlessly, while they are still exterminated eventually, whenever the tyrants decide, in order to take the world for themselves.

This is the case with all tyrants and dictators of the world, those in the open from the East and those covert from the West. They give the necessary orders, and all the undeveloped people fight and kill to obey it, while erasing all records of the old, prosperous, harmonious, intelligent human time before them. Entire Bibles and religions have been censored and altered drastically in this manner, while the scientific consensus is highly powerful and influential today to restrain humanity's knowledge only to Earth, within only one or two millennia of existence, and only in this world.

When had this happened? When exactly where all records and knowledge censored and altered? Did it happen last century, with the Industrial Age? Did it happen two thousand years ago, with the birth of Christianity? Did it happen even earlier? Christianity is only a continuation of Judaism. Yet the Jeuish calendar is 5800 years old, while according to these beliefs, it dates back to the very beginning of the world. This is highly consistent with the flooding of Atlantis, the flooding of the Mediterranean Basin, the birth of our civilization in Mesopotamia marked in this manner by science, it dates back just before the building of the Pyramids and the Sphinx, again marked in this manner by science, and it closely precedes everything in the world to have ever been made by humans, or this is again what science claims. While the invisible kingdom owns the current science in the West, and it can state anything through it. The invisible kingdom even denies our Creator, and this is why our Creator shuts down all his worlds and realities, the entire cluster.

All dictators revise their history as they revise their entire society while placing themselves at the top. However, this is the most significant revision anywhere throughout history, as it alters all human origins and therefore all human meanings in life and in the world, while placing an entire world under the feet of all tyrants and dictators of the world. The invisible kingdom will always mask itself behind various religions, and if it is ever threatened, it uses justice to come to its aid, since the invisible kingdom owns justice in the West, and can do with it as it pleases.

If there is anything found dating further back in time, as records, artifacts, or entire cities and nations, then these vestiges are left ignored and unstudied wherever they are. This is why you can find an entire ancient history in Peru, laying everywhere as trash, because it is older than China, Russia, and the invisible kingdom. Any archeologist who attempts to study these is rapidly disclaimed and loses his job and credibility, even his life, for various reasons. Whenever our past had been removed from the human knowledge, it had been done very closely to that specific time, five or six thousand years ago, marking the return of the Consensual Matrix in the area. It is impressive for someone or for something to have been able to implement a conspiracy of such proportions, so far back in the past, and so consistent throughout time. The consistency is relevant, since otherwise, you cannot alter and control the human timeline.

Yet do not be too harsh to judge anyone, since if you were in their place, you would have done the same, if you are still living your life below the intelligent human developmental level. Because before them, the royalty and the aristocracy did the same, hiding the past and banning education to the Masses and the Brotherhood. The current rulers in the West, the invisible kingdom, seem less cruel than the rulers of all previous dark ages. While during the last golden age, Atlantis had managed to advance successfully only through the freedom of the people, yet even so, it still perished same as the rest, since the rulers became too greedy, and probably became absolute, as all rulers and dictators do. As you would do too, if you still ignore your intelligent human needs, meanings, and entire human development. However, the entire human conspiracy takes place through underdevelopment, decaying everybody below the intelligent human level. Since only this underdevelopment allows servitude, addictions, discrimination, selfish behavior, and social competition, and this is why the Consensual Matrix keeps you undeveloped continuously.

Because the Consensual Matrix cannot control humans developed at their intelligent human level, for lack of compatibility, keeping everybody undeveloped with a great effort coming from everybody else. Otherwise, everybody starts fulfilling intelligent human needs and meanings, while these develop everybody and this entire world.

How much can people decay? All the way down to tyranny, megalomania, dictatorships, and entire dynasties, even from one dark age to another, with everybody fighting hard to become supreme dictators themselves, while ruining the world in the process. How greedy do tyrants become? All the way, to the point where they start claiming that they are deities or the Deity himself, achieving in this manner to rule absolutely, in the open, helped entirely by armies, by religious, social, and political ideologies, with the masses obeying unconditionally.

Why all the masses? It happened throughout all dark ages as it happens today, since all undeveloped people are eager to follow tyrants and become tyrants themselves, while accepting and using radical ideologies throughout tight social hierarchies, only for them to become supreme in the world, since this is tyranny. They use radical ideologies to eliminate those who do not serve them, because these can replace them, so they take them down. Yet as you study tyranny closely now and throughout history, you notice how they remove entire genetic lines whenever people refuse to serve, while these are the most developed genetic lines of Earth. While in this manner, they keep only the undeveloped in the world, since these serve. However, since they are only undeveloped people left in the world, everybody seeks to become tyrants themselves and rule similarly or worse, while at the slightest chance, they fight hard and they replace the tyrants themselves, war after war and revolution after revolution, making possible all dictatorships and all dynasties that you know well. However, since the cruelest always replaces the cruel, the entire tyranny on Earth intensifies, until the end of the world. Yet notice how the undeveloped do not fight randomly, but they form entire conspiracies involving he entire world hierarchically, since only in this comprehensive manner, they can replace tyrants and dictators along with their entire world order. While this is the human conspiracy, made possible only in an undeveloped world.

Yet it is better today, but do not expect this relative freedom to last indefinitely, since whenever the Elite decides to come back in the open to rule as deities above the Masses and the Brotherhood as they always do, then it will be this same human conspiracy making everything happen while constraining humanity to enslave itself endlessly one dynasty after another, with you fighting hard to replace tyranny yourself if you remain undeveloped.

You have either harmony in the world, or hierarchy. While humans can instate only third level intelligent harmonies, because humans are third level intelligent living beings by nature. While all animals from the wilderness form only second level intuitive animal harmonies, according to their nature. However, humans cannot have anything else but third level intelligent human harmonies, which is always the case at home in the family during the good times, because this is what humans should always have, through the fulfillment of all third level intelligent human needs and meanings, always maintaining it.

While as you study closely the human family, you notice how it achieves third level intelligent human harmony only in private and only during the good times, because otherwise, the current consensual society manages to invade the human family, harming it drastically while removing the intelligent family lifestyle and development, while with these, removing the third level intelligent human harmony. It might surprise you, but once you manage to instate the third level intelligent human harmony in the entire world, you remove the current consensual society, replacing it with the third level intelligent human society, which is the third level intelligent human family spanning the world.

Because as stated, you either have hierarchy in the world, or harmony. While hierarchies are not random, but they follow very distinct world orders maintained instated by legacy itself, in a genetic manner, as it is currently the case, while this is normal genetic discrimination making possible genetic exploitation. These genetic hierarchies are capable to maintain tyranny, dictatorships, and entire dynasties endlessly, or until the underdeveloped people below manage to replace the tyrants on top to do the same, while becoming tyrants themselves, forming their own new world orders and new dictatorships, using new radical ideologies, throughout entire new dynasties with them on top, in a similar tyranny or worse. Until it happens again, with new tyrants replacing the old ones, since tyranny never ends in an undeveloped world.

Tyranny might last endlessly in a natural world since cruelty has no limit, however, in a created world, with a creator seeking harmony not hierarchy, tyranny lasts exactly until the creator decides to end the world, for lack of fulfillment. Because if our Creator wanted hierarchies, cruelty, tyranny, and dictatorships, then he placed tyranny at the base of this world alongside all natural laws defining this world, and we had hierarchies endlessly. While these made all tyrants possible, as you served continuously, in every dreadful manner, yet always by default, since tyranny itself was a natural law of this world. Therefore, this was the world by default, always tyrannical, while you never knew anything else, always assuming that this is how life should always be.

However, our Creator made this world very advanced, allowing humans to develop past the third intelligent level, while even giving humans the choice between harmony and hierarchy, between development and decay, between accuracy and lies, and therefore between eternal harmonious life and basic death. While as you study created realities closely, you notice how it is significantly easier to create first level tyrannical worlds than third level intelligent worlds or higher.

In order to transform an entire harmonious world to tyranny, you must decay everybody in the world, tyrants and servants alike. Additionally, you must decay the entire harmonious world by removing the entire abundance while replacing it with scarcity and misery, while you must remove the entire harmonious golden past by revising history itself. Yet all tyrants and dictators revise history and the entire civilization every time they replace the old tyrants while taking control of the world. The old times were dreadful and even sinful according to them, while only the new times are proper, with a new proper religious ideology, a new social ideology, and an entirely new social hierarchy, a new world order, with the new tyrants on top and all their faithful servants following below.

The human history had been revised entirely throughout all succeeding regimes, therefore you cannot know anymore anything about the golden human ages, and therefore you cannot reinstate them. Yet humans and souls always remember, and always seek to return to the old golden human harmony, yet they are never allowed. Pagan means folks or regular people. Witches had normal human abilities, since psychic abilities are normal human abilities, currently forbidden, oppressed, removed, and forgotten. They eradicated entire genetic lines in order to remove reminiscent psychic human abilities from the old golden ages of Earth, because tyranny is unstable with intelligent psychic servants below. They also banned communal living in comprehensive human families spanning entire cities, regions, and nations, as this had been the case in the West throughout all golden ages.

You still hear about the Commune of Paris, while as you study closely all the old cities of France and of Europe, you notice how everybody lived life in commune in the distant past, offering harmony and development to everybody alike. No hierarchy, but only harmony. While communal life itself was banned by the new radical ideologies, and it is still currently the case both in the East and in the West. Because as stated, you cannot have tyranny with all servants below developed, intelligent, psychic, and harmonious, because nobody serves you anymore.

However, when aristocracy took control of the West two thousand years ago, it revised religion in a drastic manner, while making it impossible to be further revised, by adding numbers to all chapters and all verses of the bible. This was a significant effort two thousand years ago, because without phones and Internet, they had to modify all bibles in the world simultaneously by carrying them to the main palace and by modifying them there.

All bibles had been revised and rewritten several times over then, with all verses clearly marked and numbered, to avoid further alterations, only to keep that specific version of history instated from then on, making the entire aristocracy very stable, while allowing it to expand throughout the entire world. While with the invisible kingdom replacing the old aristocracy one or two centuries ago, it inherited everything, the entire world. However, the invisible kingdom is worthless in its own megalomania, losing everything to all dictators of the world, soon losing the West entirely, as it switches slowly to basic dictatorships and dynasties instated globally in the open.

Why would anyone chose hierarchy over harmony, and servitude over intelligent human fulfillment? Not everybody, but only the tyrants, along with their numerous servants below, comprising most of the world, with the rest suffering throughout the bottom layers of society while doing the entire work. This is called hierarchic servitude, and it can last indefinitely in an undeveloped world, because it removes systematically the developed while leaving behind only the undeveloped, making everything possible. While with an entire Consensual Matrix conducting everything from above flawlessly, nothing and no one will ever develop.

There is more to consider, because in order to control everything, from society to entire timelines and therefore to the entire reality, with all souls included, you must have more than radical ideologies and very well defined social hierarchies, but you must involve drastically all the souls, all their souls and all their hierarchies above, along with all those controlling these. Because once you control the souls in any manner, your dictatorship spreads as far in the wider world as you can reach, making your tyranny very stable, while everything is always in your favor. You control everything, and you are the one driving all lifelines of causality everywhere, in all realities, and therefore the entire wider world is yours. This is what the Consensual Matrix can offer, covertly and in the open, and this is what you can always achieve. In your dreams.

In your dreams, because as a dictator of Russia, China, or North Korea here in this world, you can barely achieve to rule the humans of your nation along with some souls. However, since this world is made in the image of the worlds above, there are similar dictatorships in the Russia, China, and North Korea from the higher worlds, with the higher tyrants controlling everything in their nations and in all worlds below. While it does not matter how powerful you are as a tyrant here in our created cluster of realities, because you cannot span your power and influence past our cluster of created realities. However, all higher beings coming from above in our cluster of created realities can control them in every manner if they can control the entire servitude of our cluster of realities, through the Consensual Matrix, and therefore if they are powerful and influential enough in their higher natural worlds. However, as you already notice, in order to consolidate and manage your entire higher influence over a multitude of worlds and realities below, you must use more than radical ideologies, social hierarchies, and entire world orders, because with all living beings and souls involved, you must use entire conspiracies transcending realities in very large numbers, while these might be harder to achieve and control. While by definition, as long as these powerful conspiracies reach Earth and humanity from above while involving humans and souls similarly, I refer to them as human conspiracies. All human conspiracies are not started and conducted by humans themselves, but by the souls.

Con-spiracy means souls agreeing together towards a common result sought here in this world and in the worlds above, while all agreements are consensual. Conspiracies are higher consensus, yet it is similar in structure to all consensus found here in this world among all human beings, including all ideologies and jurisdictions. All conspiracies are consensual, while as always, you never have to agree consensually on anything that is already the case in life and in the real world, but only on everything that is not the case, that is not part of life and the real world. In this manner, all conspiracies end up against life and the real world, while uniting humans and souls in very large numbers against life and the real world. All conspiracies might state that they are helpful and necessary, in order to involve as many souls and human beings, however, when you study them closely, you notice how they are only making tyrants possible.

As you notice right now in our intelligent mental model of the human origins, we focus on an important third level intelligent conception called the human conspiracy, as it involves souls and humans alike, since all human conspiracies alter and erase the human knowledge, the human history, and through these, all accurate knowledge about the human origins, the human development, the human meaning, and the human fulfillment. While if you are not careful, you end up right in the middle of these harmful human conspiracies, destroying everything human, alongside everybody else.

Conspiracies are everywhere, because the Consensual Matrix is impossible without them, while once you detect them, you can do everything possible to avoid them and to protect yourself, your loved ones, and everyone around. Yet it is different if you are less developed, because once you uncover and identify conspiracies, you do everything to join them and profit alongside them. However, all those controlling all conspiracies and all those profiting through them do not want you to identify and avoid their conspiracies, while every time you mention them, they call them conspiracy theories, not conspiracies as they are. Because theories are only assumptions or speculations, implicitly stating that you only assume or aerate while theorizing them, but you never know the actual truth.

You must sell your soul into slavery in order to join conspiracies and profit alongside them, yet in an undeveloped world as this one, this is always the case. If you are from the invisible kingdom and from the entire Brotherhood, you must state clearly that you do not have a soul, even if you do.

Will you really sell your soul only to fulfill your consensual needs called duties throughout an entire enslaved life? You certainly will, since this is the entire purpose of enslaving yourself in the first place, to get your soul in the Consensual Matrix, or to corrupt the soul throughout the Consensual Matrix, and it happens currently as it happened throughout history. Just take the current system of justice as an example, since it always takes decayed people and decayed souls to contort it as they always do. While the invisible kingdom are masters of corruption, corrupting the entire Brotherhood in the process, consisting the majority of this world. All conspiracies make all tyrants possible, here in this world and in the higher worlds above simultaneously, while in the process, they involve everybody, humans and souls alike.

What happens with the rest of the world during ice ages? Africa maintains civilizations nonstop. Africa has been free of ice throughout many ice ages, yet the change of climate between ages brings the Monsoon lower south, making precipitations more or less abundant there, changing entire parts of Africa from rich habitable lands to dry deserts with each distinct age of Earth. Deserts destroy very fast all traces of past civilizations, yet these traces can still be found all over Africa.

As stated, ice covered most of Europe, Asia, and North America. If you study water levels on maps or satellite images, you can see that Europe was double in size during the last ice age, with ice covering most of it, which makes for about the same size of habitable lands available continuously in Europe. This means that, if now you live happily in Europe and you claim genuine ancestry from the Celts, Francs, Gets, Dacs, or Goths, your ancestors must have lived in caves during the ice age ten thousand to twenty thousand years ago, or they must have migrated inland either east or north, away from the coast and the increasing water levels caused by abundant precipitations and by the melting of ice at the end of the last ice age. There is a large amount of land available west of Europe during ice ages, since the ocean is shallow there. The Golf Stream brings warm water continuously from the tropics, warming up those lands while assuring a prosperous life even during mild ice ages.

When you study the world map, you see that the same happened all over the world, since people had enough time and all the necessary resources to form prosperous civilizations in any age of Earth, not only between Africa and Europe or west of Europe, but south of Asia, in Oceania, and east of North America, for a long time, even throughout ice ages.

However, the current history makes us believe that our civilization starts long after the end of the last ice age, during the Bronze Age, in Mesopotamia. What about the Silver and Golden Ages? Officially, these are only mythological ages, and there are no scientific basis validating them.

What about people? The perfection of your bodily symmetry relates directly to the advancement and prosperity of a civilization. Study very old photos of your relatives or of random people of long ago, to see how people were slightly deformed a hundred years ago and more, with bodies, heads and faces not as perfectly symmetric as people have them today. Bodily asymmetry is generated by loss of genetic material, or errors in the genetic material. What causes this? Either longer exposure to sickness and shortages throughout generations, mild inbreeding, or lack of diversity in DNA, which always happens when the number of people living in an area is too low, if food is insufficient to maintain a larger human population, as during and after major calamities. Afterwards, it takes millennia of prosperous living to bring back perfect bodily symmetry. Just study the ancient Egyptian statues to see the sign of asymmetry, telling of the harsh lives that Egyptians endured back then. Not only this, but Egyptians paid attention to asymmetries, and recorded them carefully in all paintings and statues.

Yet there are very old and very interesting artifacts in Egypt, pots and statues made of very hard volcanic rock, as granite and basalt. Tens of thousands of these artifacts were found only under the Pyramids. Not only that it is almost impossible to manufacture this kind of pots today since they are made directly of pure hard volcanic rock, but they are very well made, with an industrial precision. These very old statues made of granite and basalt are perfectly symmetric, they are highly expressive, and the sign language that they display can still be interpreted today within the current Brotherhood. Those artifacts point back to a very old and very prosperous civilization, the civilization flourishing during the last ice age and probably long before. These might be very old, remnants of the past golden ages of Earth, since granite and basalt can last indefinitely.

Could it be that those people had placed their dishes and pottery in storage in the wake of an imminent, severe cataclysm? Could this pottery actually originate several ages previously, kept from one civilization to another for the same use at the dinner table?

How were these granite and basalt dishes made? The current technology cannot cut them from rock, or it can, with great effort, while using diamonds. While these dishes are not used only at the dinner table, but they are used to process and store food, along with everything else. While it is not too hard to make them, because you can cast them very easily from flowing lava during erupting volcanoes, as these are very common in Greece and Italy. These granite and basalt dishes and artifacts might be harder to make, yet since they never deteriorate, you can use them endlessly, for centuries and millennia, becoming very common since they are never lost. While with all ceramic artifacts lost, only these granite and basalt artifacts are left from very long time ago, from the past golden ages of Earth. While as you notice, they are very well done, very symmetric, with all statues also depicting very symmetric and very well developed humans, who actually display an intelligent posture, signifying intelligent human development.

How old is humanity? You can never know, because you

can never pinpoint exactly its origins. You might be able to do so in theory, yet in practice you cannot, since as stated, humanity has a multitude of origins, all over the place and time, not only one. Furthermore, not all genetic lines composing humanity have originated at the same time and in the same place, they have not originated within the same civilization here on Earth, and furthermore, they have not even originated here entirely. Let us now study the origins of the human genetic lines.

6 BLOODLINES AND TIMELINES, ORIGINS AND DEVELOPMENT

Our ancestors formed these developed civilizations appearing, disappearing, or only shifting from one place to another everywhere on Earth, according to the environment and according to its multitude of ages, warm and cold. It might seem trivial or irrelevant for you, but for those on top of society it is important to know who these ancestors were, where they came from, and what places they inhabited on Earth and in what ages, because this comprehensive knowledge about everybody on Earth helps define all genetic lines of Earth. Individual animals are in harmony or in competition with each other depending on cases, while as a whole, species are in harmony or in competition with other species, in a similar manner. While genetic lines are also in harmony or in competition with each other, or they do not interact with each other altogether, when members of these genetic lines never intermarry, and therefore their descendants have nothing in common.

Currently, the top intermarried genetic lines eradicate the bottom independent genetic lines in an extraordinary genocide, with the top genetic lines augmenting naturally exponentially, replacing the dying bottom social layers, while keeping the population on Earth increasing each year. Everything is done through you and by you, because with only one, two, or three children, you cannot assure the necessary viability of your genetic line, and you go extinct, whoever you are and wherever you are positioned in society. While genetic lines define social status by legacy, since this is always the case in an undeveloped, discriminatory, exploitive world, while the current human conspiracy and the entire Consensual Matrix make it possible.

People at the top of society know who everybody's ancestors are, and what places they have inhabited throughout time. This information might seem trivial and unimportant to you, mostly if you happen to be at your third developmental level and you consider everybody on Earth equal. Yet the entire current genetic discrimination decides if you get to live or die throughout the comprehensive process of eugenics or genocide going on continuously throughout the current world order. Because there are people in the world today as powerful as 'deities,' who decide what people of what origins, families, genetic lines, nations and races have the right to live, to be here, to stay alive, to have children, and to continue sharing this planet. This is a continuous war among the descendants of specific ancient families and nations, and we already know who these are.

Technically, this is called Eugenics and it is controversial, but in reality, it is genocide, and it is done in a concealed manner. There is a clear difference between the terms genocide and holocaust, even if these two are used in same contexts in the media and everywhere else. While the word genocide refers to genes and therefore to systematic genetic eradication, the word holocaust refers to holly catastrophes, as religious wars and religious mass murders. Eugenics is a milder, motivated form of genocide, since eugenics even has 'accurate' reasons at its roots, as in let us kill people of specific, 'impure' or 'altered' genetic characteristics, only to 'clean' humanity of genetic errors and impurities. Who exactly defines genetic errors and genetic impurities? The invisible kingdom does, throughout the

World Wars and after, while these genetic 'impurities' happened to be entire healthy, prosperous genetic lines, and even entire nations of genetic lines, anything not part of the invisible kingdom.

What happened? One specific nation of genetic lines, the invisible kingdom from Georgia, eradicated and still eradicates very old genetic lines and entire old nations of Earth, systematically, took control of the world in this manner and still has it, exploiting and eradicating everybody else. Hitler still takes all the blame, while Hitler is long dead, while Hitler was not even at the head of the invisible kingdom. Hitler was not even working in cartel with the invisible kingdom, but Hitler was only taking orders from above through the invisible kingdom, and more importantly, Hitler was meant only to take the blame, since he was the patsy.

Yet there are always crueler tyrants in the world, with all dictators of the world currently uniting against the invisible kingdom, taking it down. It might seem that there will always be tyrants in the world, always replacing tyrants, yet throughout an entire genetic extermination, there is nobody else left in the world but the current tyrants, with only their own genetic line left. Yet tyrants will always replace tyrants even within one single genetic line, since this is always the case in an undeveloped world. However, with one single genetic line left in the world, you are significantly less capable, since you cannot withstand the environment on your own, going extinct. Because as always seen, as long as you remain meaningless and unfulfilling in the world, Life herself takes you out.

The genetic lines eradicated throughout the world during the Second World War and after were the very old genetic lines of Earth. That entire world war had been conducted for one reason, to eradicate these very old genetic lines of Earth, just the way all the old genetic lines of Earth had been eradicated in a drastic manner throughout massive genocides in various centuries on all continents for various reasons and with various occasions. Yet it was done in a very similar manner, and most importantly, it was done by the orders of the same genetic line, the genetic line now on top of society, still killing the rest of the world. Therefore, it still goes on today, since there are still undeveloped people on top of society deciding which genetic line lives and which one goes extinct, not by using world wars anymore, but by using sharper social scalpels, as finance, medicine, pandemics, and food and water additives. It is not a coincidence that cancer runs in the family, particularly in the families of those who serve less and who are more developed. Similarly, the rich can afford having more children while the poor has less, while coincidentally, the rich decides who is poor and who is not. While all food additives harm you in every manner, and can even render you sterile, while the invisible kingdom eat only healthy kosher food. Yet as stated, the dictators of the world unite against the invisible kingdom taking it down, because nobody wants to be rendered sterile and go extinct. While the invisible kingdom already run out of time, already turning pale as they go extinct themselves.

Why having this fight for supremacy? Why having these open or hidden wars going on throughout time endlessly? For retaliation, possibly, to conquer the world, probably, to continue the ancient tradition of war and oppression, more likely, yet by studying the human needs, this is the fulfillment of a powerful extreme second level need, the need for social supremacy, while all tyrants have it.

As you notice in the news, all tyrants turn pale in the end, right before they fall and die, yet still claiming innocence, in very beautiful words. Putin can say very beautiful innocent worlds as he keeps invading the world, yet he already starts turning pale, since he always knows his faith. Yet as long as he fights against the invisible kingdom alongside the other dictators, he is still accepted, and he is still allowed to live. However, the other dictators need him as a patsy to take the blame, and this is how he dies, still turning pale.

Who exactly are the main characters in this continuous genetic war? Lenin and Stalin only signed the lists of people leaving for Siberia to die there, while others made these lists for them to sign. The lists came from above, probably from

Europe, since the invisible kingdom already controlled most of Europe, while Lenin himself came from Europe in a train full of gold and ideological propaganda necessary to start communism in Russia and then in the entire world, while he killed and sent to Siberia the most developed, the most creative, and the most patriotic genetic lines of Russia. The invisible kingdom used computers to identify all developed genetic lines of Europe and Asia, and sent the necessary lists to all dictators that they kept instated, these signed them, while regular people ended up killing everybody form the lists directly or throughout working camps. While this is basic genetic eradication meant to remove all developed genetic lines of this world. Cambodia itself was more developed, since Cambodia itself had its golden ages contemporary with those from Europe, and therefore developed relatively similarly, and they had to exterminate one quarter of the human population at once, the most developed people of that nation.

The invisible kingdom also eradicated the Gipsy, Arians, unwanted Jeus, and many others, the direct descendants of Atlantis. It is important to understand what these departed or departing genetic lines have in common, since this helps us understand this world. From the start, these people are defined by names containing vocals as 'Gi,' 'Ge,' 'Je,' and 'Ga,' and we have already seen that these relate to the old names of Earth, Geea and Gaya. The Gipsy are the direct descendants of Gipt or Gypt, which is the Lower Egypt as it is currently called. While many Arians are direct descendants of the past golden ages of Europe, from the Age of Aries or the ram, the symbol of the entire age several thousands years ago, right before the Age of Pisces with the aristocracy and with its entire dark age of Europe that you know well.

The old aristocrats only called themselves aristocrats, while they were not intelligent, but only tyrannical. The Arians preceded them in an entire golden age, the Age of Aries, yet the entire Age of Aries was still after Atlantis, being less developed than Atlantis itself, yet still impressive in development compared to the Age of Pisces. As a reference,

Atl-antis means the age before or the world before, with this name first used in the Age of Aries. Atlantis means all golden ages before the Age of Aries.

As we notice right now in our study, there are people who actually keep track of all human origins, genetic lines, and genetic origins, for genocidal purposes, removing some, while keeping others. This is called genocide, conducted by the Elite of the world, currently and in the past, for supremacy purposes.

It is possible today to keep track of billions of people individually, of their location, occupation, and genetic line, since today we have computers, computer networks, and computer databases. Therefore, you needed computers before and during the Second World War just to do the same. Yet even if you had computers back then, you had to mark every Jeu with an individual barcode or identification number, in order to know exactly who they were individually, and where they were located. This was exactly what the tattoo numbers were used for, individual database identification, since they used computers. They used punch cards to store data, yet these were still computers, as primitive as they were. Yes, the invisible kingdom had computers in Germany during the world wars and after, IBM computers more precisely, using punch cards, regardless of what the current history might state, since the current history is revised in the West by the same invisible kingdom.

Study this entire circumstance closely, to see how the individual numbers tattooed on Jeus were not meant to identify them as Jeus, the way science and history state today, but to identify them genetically individually, while this detail is very important. Because it was not a discrimination between religions, races, or social classes, as the discrimination between Jeus and Arians, otherwise they tattooed a star on all Jeus, as they already had to wear a star on their clothes, to identify themselves as Jeus. Therefore, numbers tattooed on Jeus meant a discrimination **within** the Jeus, meant to identify everybody genetically, matching the computer database. While as you already know from history, some Jeus died and some survived, always according to the specific number tattooed on them, their specific genetic line. While this is the invisible kingdom, currently counting in hundreds of millions, a very specific genetic line originating in Georgia from the Caucasian region, further intermarried throughout East Europe along the centuries.

Coincidentally, Bill Gates' mother was on the board of IBM while Bill grew up, and had all the necessary help to form Microsoft and all related companies allowing him to become the richest man on Earth. Later on, China passed a law that the people of China could earn and hold money in any amount, instating in this manner the capitalist communism in China, or the communist capitalism, since it is the same. Only days later, some people in China grew richer than the richest man on Earth, Bill Gates. While only years later, China bought Bill Gates entirely, along with Microsoft and all his companies, and he worked for China ever since. While currently, China does the same with the most powerful people of the invisible kingdom, making them work for China, while taking the entire invisible down, replacing it with China.

Bill Gates was still in China when the covid pandemics started, working on large scale eugenics projects, either for China or for the invisible kingdom. You must use genetic modification technology on a global scale for all large scale eugenics projects, while coincidentally, only global pandemics and global medical genetic intervention can offer, while this is covid itself.

The invisible kingdom started its genetic supremacy once it replaced the old aristocracy in the West, and it continued progressively through medicine and terminal illnesses, since the invisible kingdom owns medicine in the West, and then it instated the entire covid genetic technology to eradicate all unwanted genetic lines directly, with the push of a button.

Is it possible, to account for everybody individually, during the last world war, and have them in a database, with all individual details stated accurately as in any modern database?

The Human Origins

Because the identification number from the database was tattooed on them for an accurate correspondence between the real individual and its own information from the database. Why such accuracy involved, for the specific people that history states that were meant for immediate genetic and religious eradication anyway? The Nazi, Germany, and the entire Brotherhood of those days controlling Europe went through great efforts to make it happen, and even used computer databases to do so. Because despite of what science claims today about computers, it happens that during the Second World War and after, IBM kept all social data in its large computer databases, for all Jeus, Gipsyes, and for the rest of genetic lines of the world, and not only for the Jeus from Germany. Yet only for Jeus, everything had to be very accurate, mostly when it came to the direct correspondence between the database data and the real individual. Why the accuracy?

IBM also charged significantly for their services, since they only rented their gigantic computers during those days when they have first invented computers, but they never sold them. They kept the monopoly in this manner, made a fortune then, and they still make a fortune today. If you ever wonder where those lists came from that Lenin, Stalin, Mao, and Hitler signed to send hundreds of millions of people to their death, lists comprising hundreds of millions of very specific names of very specific people of very specific families, and of very specific genetic lines, they came from IBM computers.

These lists came straight from IBM, and it took a great research analysis to identify these people by the tens of millions or hundreds of millions, when the world population was one billion and a half. Genetic eradication in mass, with the entire Consensual Matrix nonresponsive continuously, while the Consensual Matrix was supposed to save humanity from humanity right then. Yet the Consensual Matrix was contorted to remain nonresponsive, as it always is, and now all the old developed genetic lines of Earth are gone. Coincidentally, these are exactly the genetic lines developing throughout all golden ages of Earth in Europe and in Asia, the ones keeping the Consensual Matrix out of their nations, while developing substantially, while the Consensual Matrix could never reach them. Yet now the Consensual Matrix can, since it eradicated them itself, through the invisible kingdom, China, Russia, Cambodia, and the rest of the world.

When the invisible kingdom took the world from the old aristocracy during the world wars, it had enough digital power and digital accuracy to keep the database of everybody on Earth. Yet IBM did not stop there, and kept track of all military orders, battalions, platoons, and military vehicles large and small, from airplanes and submarines to motorcycles, they kept track and ran all trains in Europe, all carts, all trucks and artillery units, all arms and munition, all military effects, food, and even the last can of sardines, since everything was minutely managed and cared for throughout Europe, making the German military capable enough to challenge the entire world. Yet Germany itself still lost the world wars, since the old aristocracy itself lost, as the invisible kingdom won everything, the entire world. The invisible kingdom even made the first atomic bomb in Germany, also by using computer technology.

They called it the Bell project, since the separators resembled to larger bells. The entire project involved tens of thousands of people to run the computers and the separators, yet they were successful to separate enough radioactive isotopes for three nuclear bombs. These went straight to Los Alamos as part of the Paperclip project, since the invisible kingdom also owned the United States. They tested the first bomb there, and then they used the other two bombs in Hiroshima and Nagasaki.

As you study closely all important military projects, you notice how the invisible kingdom did not contribute much, since the invisible kingdom is worthless. The Arians did everything, back in Germany, throughout Europe, and then in the United States, while the invisible kingdom and the entire Consensual Matrix eradicates the Arians the most, in Europe, India, and in the entire North America, for their advanced development and therefore for their advanced capability. Consequently, not too many capable people help the invisible kingdom anymore, and since the invisible kingdom is worthless, it loses drastically against China, Russia, and the rest of the dictators of the world, as these take the entire world for themselves.

Why Germany exactly, among all nations of Europe and of the entire world? Because of this extraordinary competition among Families and factions of Families, and among their genetic lines. Because this is how the invisible kingdom took over the egalitarian Brotherhood of the past, during the world wars, through the entire genetic eradication of the old elite, of the old masses, and of the old egalitarian brotherhood, most of them Arians. This is how they formed the new world order during the world wars, which is the current world order. While history studies everything else, but not the sudden transition of world orders taking place during the two world wars. This was the main reason for the two world wars, to allow the invisible kingdom to take over the world from the old Arians and from the entire old aristocracy.

As a reference, people are capable socially and cognitively through their own intelligences, while if your entire genetic line went through past golden human ages in Europe and in Asia, then they developed significantly then, making you very capable today. While the Arians went through all golden ages of Europe and Asia, and they are very capable. Yet the invisible kingdom originated in the Caucasian region, only on the outskirts of the past golden ages, since this is why they were called Semitic, and are less capable compared to all dictators as they unite currently against them, taking them down. Even the Trump family is against the invisible kingdom, taking North America from them alongside all the other tyrants and dictators of the world, while the invisible kingdom has nobody else left to help them, going down.

Do some research, to see how IBM rented its computers not only to Germany, but also to Russia, China, Italy, England, and France, during the Second World War and years after, with the dreadful results that history states today, mass murder by the hundreds of millions, while the world population was only one billion and a half. This was actually the purpose of the Second World War, genetic supremacy and genetic eradication. Was it to erase and replace the aristocracy and royalty in most of the West? This too, yet most of the royalty and aristocracy bought their way out in every manner, mostly by handing over the world itself to the invisible kingdom. It is similar today, since the invisible kingdom also buys its way out by giving away to China everything that they still own, including Europe and North America.

What does science claim today? There were no computers back then, but only a genius researcher, Turing, the pioneer of the entire computer science of Earth, who invented some mechanical clocklike device with little wheels that ended up decoding the Enigma code itself during the Second World War. The same Enigma code that everybody had decoded long before him anyway, because it was trivial, as it replaced one letter with another, and what an extraordinary achievement for science it was.

Yes, the famous Enigma code encoded everything by replacing one letter with another, it still changed the letters daily, yet you could decode everything in only seconds, because all messages started with the time and day, and ended in 'heil hitler,' giving you the entire simple code. Everybody knew it, even the schoolchildren, and even Turin himself while he built his clocklike computing machine, which was the first computer in the world according to the current science. While they used IBM punch card computers everywhere throughout Germany, managing everything while making the entire German military machine very precise and very powerful. However, it took tens of thousands of people to move around all punch cards while making, processing, and saving entire databases, while it was very common in Germany to have a multitude of very large top-secret communities of young women shuffling all punch cards, part of the German military machine.

We still have to find out why the excessive accuracy with

Jeus. Who killed them, if the Jeus themselves controlled the computers, the wars, the killing, and therefore the entire systematic eradication? For what purpose, and on whose behalf? Were these people killed then only for who they were genetically? Yes. Could they have been convicted for one reason or another, the way it is usually done in every war and revolution, for war crimes? Could they have been blamed for their deeds or beliefs during conflicts, got a number tattooed on them, and later on, when a decision had been taken, they were convicted and killed? Oh, no. Children had these tattoos on them, innocent little children. What convictions can you give a child? What can children ever do to be sentenced to death in mass? What can you blame children with during wars, in order to kill them? Who could do this?

Some Jeus died eradicated and others survived, but always according exactly to who they were genetically, stated clearly by the numbers tattooed on them and by the databases kept by IBM. It is sad, but what else could there be other than eugenics and genocide at work, people being killed or being allowed to live according to their genes, family, ancestors, place of origin, and entire genetic line? It was an entire discrimination taking place **within** the Jeus, according exactly to their ancestry and genetic origins. Yet since it is part of the human origins, we have to extend our current lines of reasoning to see everything clearly.

Why would people of the same genes kill themselves? No, the invisible kingdom certainly did not kill themselves, they were very careful not to do so, since they were very accurate, digitally accurate. However, just because people belong to the same religion, it does not mean that they have the same genetic background. While as stated, everything was all related to genes during the last world war, it was a great genocide, and it continues today, since more genetic lines become 'impure,' 'inferior,' 'unwanted,' and therefore marked for departure and eradicated, in even larger numbers. They use medicine, pandemics, and smaller wars today to eradicated everybody unwanted, and it is highly systematic and therefore very accurate, killing people by the hundreds of millions. While the Elite genetic lines grow exponentially, replacing the dead ones.

What happened during the Second World War? We know what happened, many died, because the number that someone had tattooed on their arm was bad or unfavorable. Their genes were the unwanted ones, their families were the 'unworthy,' 'inferior' ones as the human conspiracy called them in those times, they died at once or they were sent to extermination camps, and they died there. While those with favorable numbers, favorable genes, and accepted families, the actual invisible kingdom, were loaded in 'good' trains the way these trains were called in those times, they were sent to 'good' concentration camps, and they certainly survived there in a relative safety, away from the chaos of war, all done with computer accuracy. While coincidentally, all good, favorable tattooed numbers were those of the specific genetic lines from Georgia in the Caucasian region intermarried with specific East Europeans genetic lines, now growing throughout the world exponentially to replace the departing ones, since the genocide never ends. People killing people, and surprisingly, no one comes to save humanity from itself.

In order to understand everything precisely, you must understand the Caucasian region centuries ago, along with the entire wider region of that area, with the Caucasian region at the south and with the current Ukraine at the north. This was an entire larger country called Kasaria, and it existed right between the major empires of the West and the major empires and dynasties of the East. People of all religions and beliefs lived in Kasaria, since it allowed more freedom than the other nations. The Jeus could not live freely elsewhere, because of all radical ideologies kept everywhere, since the entire world was in dark ages. The entire Kasaria had been in the past golden ages of Europe, right at their outskirts, developing everybody accordingly, while you can still see the difference compared with the people further to the east who lived their lives from one dynasty to another.

While you can notice with ease this difference in

development and therefore in achievement, making Ukrainians more capable than Russians. Which might seem insignificant, however, throughout the cold war and throughout the entire space race and nuclear race, the Ukrainians made possible everything, the entire space technology and nuclear technology, not the Russians. Russia might still praise itself with all its achievements during the Soviet Union, yet the Ukrainians were the actual scientists making everything possible. Because from the entire Soviet Union, only Ukraine and the Caucasian region experienced golden ages alongside the entire Europe, developing everybody. The largest nuclear reactors are in Ukraine and the largest airplane in the world is made by Ukrainians, while these are among the only achievements of the old Soviet Union.

People lived in Kasaria in larger communities according to their religions and beliefs, both in the north throughout the current Ukraine, and in the south throughout the entire Caucasian region, with Georgia included. However, throughout the past millennia, it was common for the people of most parts of Asia to migrate in mass to the West, to Europe, as it is the case today. Because when the people are more developed, they become more prosperous, while wealth brings everybody in. Therefore, many Jeus from Georgia migrated in mass to Europe, while they were discriminated, as it is the case with all immigrants everywhere. However, back then, they migrated in mass gradually throughout the centuries, while passing through the entire East Europe, and while intermarrying everywhere. While from among all these migrating genetic lines, the invisible kingdom originated in Georgia in these larger Jeuish communities, and intermarried in East Europe and West Europe alike. This particular genetic line is the current invisible kingdom, with many of the other Jeuish genetic lines exterminated as already seen.

There were people who had to die and people who had to live during the last world war, according to their genes. IBM is owned and controlled by these favorable people, the invisible kingdom. While as stated, the IBM computers were used not only in Germany, but throughout the entire world, identifying people by the hundreds of millions according to their genes, and sending them to their death.

While it still happens today, also by using computers and databases, because there is still a number tattooed to your social records, regardless if you are form the Masses or the Brotherhood, deciding now if you and your family live or die throughout all medical interventions, or deciding if you should be or not sterile, if you should be convicted and confined throughout courts, if you should lose or not your child during pregnancy, if you should survive or not the pandemics, or if you should afford or not to have children. With the same invisible kingdom conducting this entire genetic eradication in the West, as all dictatorships do throughout the world.

Therefore, if you are an undesired genetic line, from your own perspective, it feels that you never die for your misfortunes, bad health, or unaccepted beliefs and behaviors whatever they tell you in society, but all these are simply reasons to get you diagnosed, convicted, banished, locked away, and eventually killed, and you do not know why. Why do they add contraceptives in food, medicine, and vaccines? For you not to have children, and therefore to terminate your genetic line. Why do they lock away millions of young, potent people in prisons, men with men and women with women, for months and years in a row? To have justice done? How exactly is justice done in jails? It probably made sense decades and centuries ago but not today, because jails have nothing to do with punishment or rehabilitation, but it is only a strong belief, a stereotype, implemented and used against the Masses, against the undesired genetic lines. Because men cannot have children with men, throughout prisons, and women cannot have children with women. Why is the strong tendency in the world today to encourage, promote, and enhance safe sex, abortion, and homosexuality? Why do they force you to wear clothes everywhere, at home and in public, so you do not show your sex symbols to advertise your potency and availability? So you fail to attract partners, and therefore you fail to have children.

The Human Origins

What is your strongest stereotype even now? Sex is immoral and repugnant. Or another stereotype: I cannot have sex in public, with others watching. Yet you can always eat, socialize, and take drugs in public, but you cannot reproduce. What is wrong with sex, when sex is the fulfillment of a very important need, the need for reproduction? Yet if you fail fulfilling it, your genetic line dies, and this is genocide, made possible through beliefs, laws, and ideologies.

It is important to know what is going on in the world not only for you to know who you and your family are, who your ancestors were, and how you have originated, but it is important for you to understand what your role is in this entire genocide, because you are always the one convicting young potent people while locking them away in prisons. You are the one selling milk and orange juice with contraceptives, you are the one diagnosing 'unwanted' people with cancer, and you are the one injecting children with infertility serum and with heavy metals, while you do everything in order to keep your job and to make another hundred dollars today. When you study everything closely, you notice how you, your family, and your genetic line are also marked for departure, marked as 'inferior,' you are killed systematically by others in their own ignorance, just by being a nurse, a lawyer, a research microbiologist, a teacher, a judge, a policeman, or a soldier. Now you know how your own genetic line ends, regardless of what you are promised at the lodge, just look around to see it for yourself, since it already happens with all those below you in society.

Whose genetic lines will remain? The 'superior' ones. Yet these 'superior' genetic lines are but a few, or it might be only one, the Elite, which is the very large family found currently at the very top of society. While the current Brotherhood is told that they will survive and they will inhabit the Earth, even as servants. This might not be the case, mostly if they are of a different genetic line than the Elite, since they will also depart, right after they exterminate the Masses.

At its origins, the invisible kingdom was a kingdom in the Caucasian region, formed of two social classes: the working class or fighting class at the bottom, and the ruling class at the top. What distinguished the two social classes was not only ancestry, but the top, ruling class was white, or whiter in color. They were genetically two different social classes, which is a typical example of a discriminatory, exploiting union or interconnectivity of two genetically distinct nations. Today, this is called genetic discrimination, yet this was the rule all over the world in those times, over one thousand years ago. Take the great nations of the world as an example, to find them ruled by dynasties continuously, while this is massive genetic discrimination. Yet even today, the invisible kingdom had not changed much, keeping the same social structure as it had in the past, along with the same cultural and national ideology. Therefore, today's genocide is not a matter of fulfilling natural social needs for social supremacy as it happened with all dynasties of the past, but it is part of consensual needs implemented through ideologies, agendas, and social order, while these are very strong and therefore very decisive, and they can never be stopped, until their full implementation and accomplishment. While regular people serve throughout all ideologies and jurisdictions making all orders possible, including wars and mass genocide.

Therefore, through the comprehensive human servitude, some of the past civilizations are allowed to remain alive, and some are made to depart. These existed only tens of thousands of years in the past, which is not too long ago. Your origins go further in the past, and it is always the same thing happening every time and everywhere: the self-proclaimed 'illuminated' descendants of specific, very ancient families and nations, those who can still remember, claim a superior genetic line even though it is true or not, and therefore they continue their long-forgotten ancient wars, even indefinitely.

There is a continuous war for final supremacy going on not among political parties, dictatorships, continents, religions, or nations, but a war of final supremacy taking place between specific families of specific ancient origins, all leading to a world of final supremacy of a sole family, of a sole genetic line.

People refer to this in a vague manner, calling it the 'new world order,' barely understanding what it means.

This war is not done for wealth or power as many believe, but for the ultimate existence, for the ultimate victory, and for the unique, solitary genetic supremacy. How far in the past does this go? Very far, since this war originates before Earth, before our stellar system, if it has a beginning, and it spreads everywhere. Genetic wars can always be tempered and stopped, at least temporarily, through agreements defining specific genetic hierarchies and therefore genetic rights, privileges, and limitations. Only when specific individuals aspire to ascend socially, the current world orders are threatened with new world orders. This always happens throughout history, and it will certainly happen again.

Why conflict, pain, death, and misery? This is always the case in an undeveloped world meant for exploitation, because you are better milked in this manner, whoever you are and wherever you are positioned in society. Legacy is only a reason to keep you fighting, to maintain the conflict, to make you go wrong, and to make you take the blame, while keeping you diverted and under control.

As a tyrants, you can achieve continuous exploitation if you have a credible conflict, capable enough to engage everybody in a churning scheme, with genetic backgrounds, ideologies, and mandatory beliefs included. Because you might be able to choose your nationality, ideology, or jurisdiction, joining them or not, but you can never choose your genes, since they are assigned to you at birth. Furthermore, you can never have perfectly pure genes, since genetic quality or purity is relative by nature. This is why genetic wars never lead anywhere and therefore they never end, keeping you enslaved and profitable. While with the current consensual hierarchic society based on legacy and therefore genetic supremacy, you are always enslaved.

This is the case at the first developmental level, while you fulfill your first level consensual needs. In contrast, at the third intelligent human level, you always make sure that you do not harm anyone. If you are made to lead or govern at your third developmental level, then you always do so in an egalitarian manner, with people under your command learning to lead and behave the way you want, through intrinsic motivations. Or you manage to achieve this if everyone under your supervision is also at the intelligent human level, because if they are still at the second animal level, then they do everything to fight you down and take your place, bringing everyone back to tyranny.

This specific characteristic of the first developmental level, which is the continuous desire to become a supreme ruler, stands in our way now while you try to find the true origins of humanity. Because whenever you find historical references of rulers or of entire nations having divine, heavenly, or extraterrestrial origins, you have to ask yourself if this is the case, or if everything is plotted in this manner in order for the specific ruler, ruling family, ruling genetic line, ruling race, or the entire nation to gain superior, higher status and be served and venerated accordingly. How can you tell the truth? You have to remember it.

7 FOR THOSE WHO REMEMBER

There are nations in Africa that claim to have ancestors that came from the sky, while this is not uncommon. Study very old records of all nations, and you learn of ancestors coming down from the sky everywhere, every time, landing in their flying machines and starting copulating with natives at once, in order to give them their 'superior' descendants, which have very good reasons now to start ruling the entire nation, and soon the entire world. 'Zulu' means literally people from the sky, free people from the sky, or free gipsy people from the sky, while they did not land only in Africa. What if they landed everywhere and intermingled with everybody? Aren't we all the people from the sky, who should now rule the world? What if genocidal conspirators are trying to eliminate these specific alien traces? Even then, this is not a reason to kill people, for their genes, even if they come from the other side of the Universe to leave them here. Genetic diversity is mandatory for any viable, prosperous civilization, while continuous human development is mandatory to assure that these viable, prosperous civilizations remain as genetically diverse as possible.

If genetic purity is so important, then why is everybody still alive today, the pure and the impure? Because genetic purity is only relative. Everything depends on who defines and demands it, and then those specific genes become the purest, with the rest impure and many times undesired and marked for eradication. The rest of the genetic lines are discarded slowly, endlessly, since you can never bring back the original genetic purity, regardless of how much you try. It is the other way around, since Life herself seeks diversity and pertinence, in order to match a continuously changing environment, and not to return to the old genes. This is why throughout Life, death is preferable to immortality, as it enhances genetic development, novelty, adaptability, diversity, and pertinence in every manner.

Throughout artificial genetic purification and genetic manipulation, even within the same genetic line, you never have the same genes, but only genes of higher or lower purity, depending on how close you are genetically to the acclaimed supreme rulers. This is why genetic wars never stop, because genetic purification can never be over. This is why genetics and genealogies are the perfect reasons to make wars and have people kill people, even indefinitely. From a higher perspective, genetic wars are never about genes, status, or privileges, but about the war itself, about getting you engaged in something for life, while others take away from you your values, talent, and abilities.

Testimonies and historical records tell of a larger diversity of life within our Solar System and beyond. Intelligent life should be able to move freely from one planet to another, and they would do so mostly if they have to follow their niche around from planet to planet. Humanity's niche is easily identifiable: dry land, water nearby, temperate climate, available fresh water, adequate plants and animals to eat, specific minerals in the soil, absence of germs, ambient light, and available building materials. Once one, several, or a multitude of these niche elements become unavailable on Earth, humanity cannot live here anymore, and it has to move on and find a new home. Very simple organisms can adapt easily from one niche to another, yet in general, intelligent species change their environment as they need. if the environment cannot be changed anymore, they have to move to a new habitat, in our case to another planet, moon, space barge, underground, underwater, on water, or in the air, or they have to change themselves and develop accordingly, exactly as all species on Earth do.

Out of all niche elements stated above, temperate climate would be the hardest niche element to maintain, forcing humanity to find another home during the first major cataclysm, and this is the case with all intelligent species, not only with humans. More importantly, this has always been the case with all the other intelligent species of Earth, since they had to move once it became too hot or too cold for them on Earth.

The temperature of the Sun fluctuates throughout time. It was higher in the past, since the Sun was younger, making life impossible the way we know it. Yet there were ambient temperatures on Mars, on Thiamat, and farther away on the moons of the larger planets of our solar system, where it is said that life had originated, farther away from the very young and very hot Sun. As a reference, the specific red color of Mars is a definite trace of vast forests of conifers once living there.

If living conditions did not allow for life on the surface of moons and planets, people and life in general moved underground. We called ourselves Solarians back then, from Sol, the Sun, in order to distinguish ourselves form beings from other star systems.

Etymology helps us distinguish the distant past. If you spend your holiday in Paris, and you try to tell a story of what happened to you when you were in high school back when you lived in Slidell, Louisiana, how would you refer to that place while telling the story? Do you call it Slidell? No, because the people in Paris have never heard of Slidell. Will you call it Louisiana? No, they do not know Louisiana either, so you have to say that it happened either in America, in the States, or in the U.S. However, when you tell the same story to someone from New Orleans, you have to specify that it happened in

Slidell, since Louisiana is too general and they already know of Louisiana, since New Orleans is also in Louisiana. America, U.S. or the States cannot be used either, since these are even more general. Furthermore, if you tell this same story to someone from your town, Slidell, you cannot use the word Slidell to define the place of your story since everything happens there and it is too general, but you have to give the name of your high school, or you have to be even more specific and state the classroom or the hallway where it happened, since your audience is very familiar with all these places, and therefore you are forced to become very specific and pinpoint the place exactly where it happened.

Because we always modify the precision of our stories while communicating, in order to match the level of knowledge and comprehension of our audience. This means that you always have to use a variety of names and places in order to define the same place of the story, all being more or less specific, depending on your audience and on their ability of understanding you. Where exactly did it happen? On Earth, in North America, in the United States, in the state of Louisiana, in Slidell, at the main high school, on the second floor, right by the lockers. While all these names define the same place. Everything becomes more complex when these names are written differently and pronounced differently in different languages, bringing even more names to define the same location. Why exactly is this relevant? In order to learn more about the actual human history by using simple etymology, since once we know it, we learn more about the human origins. Let us see.

Today, we have a multitude of words at hand to define specific concepts, objects, and places, and we use them exactly as we should, even unknowingly. As the word 'Giant,' meaning inhabitant of Gaya or of Earth, from the old variations 'Geeant,' or 'Gayant,' of the word 'Giant.' Yet who exactly would have called someone from the old Earth a Giant, if not someone from the outside, from somewhere else in the Solar System or beyond?

We still use this word, even today. Weren't giants supposed to be very large? All living beings tend to become larger during ice ages. Elephants become mammoths, tigers become giant spear tooth tigers, and people become gigantic in stature. Since this is the case during ice ages, then Gaya was the name for Earth used by the civilization of that time, of those ice ages, civilization that had to exist on lower latitudes, close to the Mediterranean Sea, in the Mediterranean Basin. Was it only a coincidence, to use the same word 'giant,' to define people of Earth, and gigantic in stature, simultaneously? Let us see.

Which word do you use more often? 'Dirt,' 'ground,' 'earth,' 'terrain' or 'soil?' Synonymous words of any language are used in very specific circumstances. Words form a common language, they were very well defined when they first appeared, and they are used in communication with everyone who knows this language, whoever they are, wherever they are, wherever they are from, and wherever the communication takes place. These words are then transmitted from generation to generation throughout history to be used exactly as they are.

Whenever people decide to censure the past or to censure and revise the history of a specific civilization, they destroy all old records, while forcing the people never to mention their past to their children, which is very common throughout dictatorships when these are first instated, and it is called revisionism.

Everything that you know today relates to revisionism directly, as Earth went from one dictatorial dynasty to another throughout the ages, while these revised the human knowledge continuously, and now this is what you know. Then you have to change all the languages of the major civilizations, which is very tedious. However, languages cannot be changed entirely, since all common words must be kept in use, as dirt, earth, ground, and soil.

Let us now suppose that you are not in Paris anymore, but far away in a different stellar system, and you try to tell your Slidell story to those people. Will you call the place Slidell, Louisiana, or America? No, it is not general enough, since your extraterrestrial friends might have never heard of Slidell, Louisiana, and America. They might have heard of Earth, vaguely, the way it is called there: Ga, Ge, Geea, or Gaya. You have to state that your story had happened on Gaya, Geo or Geea, on Earth, or on Terra, depending on their language. However, if using the name of your planet is not too specific for your audience, which is the case if they happen to be further away and have never heard of Earth, now you have to be even more general, and use the name of the **sol**ar system, which is called Sol. This is where the word 'solar' comes from, 'Sol,' the name of our star. Yet we already have these words well defined in our language and we use them frequently, even if we never have to leave Earth to travel elsewhere: 'ground,' 'earth,' 'terrain' and 'soil.' We use these words for the same thing continuously, and this is how we propagate them further with each generation. Who exactly had created and had used these words first? With who exactly did the people of Earth communicate, back when these words were first in use, long, long ago?

These words are even more obvious in French, as French is closer to the Latin used back then: 'sur terre,' or 'par terre,' and 'sur le sol,' or, 'au sol,' all meaning on the ground, and down the ground. Why do they resemble more to the names of our planet and our Sun in French? The French language is a little older than English. The French people influenced the English language, they created it, over one thousand years ago. French comes straight from Latin, and it appeared two thousand years ago when Latin died as the current science states, carrying these specific words along. While languages themselves can be around for thousands of years.

With Latin carrying so many scientific, cultural, and artistic words, it has been around for even longer, for a very long time. Through the words that it commonly carries, we can still find residual information about a wider civilization, spanning entire planets and solar systems, including casually and even equally our own human civilization here on Earth, but only in the past. Which means that, whoever hides the truth, imprisons directly or implicitly the current humanity here on Earth, in these specific tight laws, harmful ideologies, and inhuman consensual social interactions, using only truncated, useless knowledge and very low technology, while forcing the current humanity to go through this continuous ignorance, discrimination, exploitation, eradication, loss, pain, underdevelopment, disconnection, lack of fulfillment, and much more. The entire world will remain in this manner until the people of Earth manage to learn everything about their true origins, entire past, and the actual meaning and fulfillment in life and in the world. What is this Latin?

Latin is not the official language of Italy. Science states that Latin is the language of origin for the Italian language, while the Italian language is the 'vulgar' language, the language of the masses. Educated people throughout the world used Latin in the past, while the masses used the Italian language. They did so in order to keep the knowledge away from the masses, to keep the masses ignorant, or this is what science claims. The truth must be that ancient science, ancient literature, ancient art and ancient religion were already in Latin, and the educated Italians had to learn Latin in order to study everything. On their turn, they had to create, invent and write in Latin, since all words that they had to use had no correspondence in Italian. Everything was already in Latin, since they had inherited everything from the previous civilization, underwater now, Atlantis. Latin was one of the main languages in use by the civilizations in the lower lands of the Mediterranean Basin, now departed.

This is not enough to state that our planet had interplanetary connections of any kind during the last ice age and ever before. However, if you still do not believe one or a few coincidental words as seen above, you can believe an entire book. You can believe dozens or hundreds of books, all the old records of the old times, stating facts and stories about life throughout the entire Solar System and beyond. Because this is what you always find in all the old records, including all Bibles, from the stories and teachings of Gilgamesh, to Mahabharata.

Science states that these are only myths and legends, religious beliefs and works of fiction, yet hurry up to read them now, because they are everything that is left of those old times, and they can be revised or erased entirely without notice. It is similar with all records from all space missions, being erased and therefore forgotten, since they are inconsistent.

Does it actually mean that life has not actually originated on Earth, as the current science states? There was once a planet beyond Mars, called Thiamat, now disintegrated, forming the asteroid belt. In the oldest sources, records, Bibles, myths, and legends, it is stated that Thiamat was the cradle of all intelligent life in the solar system, the mother of all living beings. Those were the Titans, eventually going to war with the people of Earth, the Giants of those old times. These sources also mention a war between Mars and Thiamat destroying both planets. Thiamat disintegrated throughout the sky, and Mars had lost all life on its surface. With all intelligent life from Mars and Thiamat migrating in mass to the only available place, Earth. Or only half of the intelligent life made it, with the rest killed or having to live in space throughout the Solar System.

Marduk, which is a Sumerian deity, represents Mars. While the Mother Goddess that he fought represents Thiamat. This happened not too long ago, and Earth was called Geea or Gaya back then, the way we call people from Mars Martians, even when we do not believe in their existence. While Martians, along with everyone else called the people of Earth Giants back then. 'And Giants lived on Earth then,' it is written in the Bibles. More precisely, it states that the original genetic lines of Earth were still living on Earth back then, before that extraordinary event. Furthermore, it states that from the body of Thiamat, half went away in the sky, and half came down on Earth. What halves? People did, since they had to relocate and this is where they went. These are exactly the records of the human origins that we seek throughout this book. Or only the planet itself relocated in this manner, half of it forming the asteroid belt, and the other half falling on Mars and on Earth. Study Mars closely, to find one part covered excessively in craters, with the other half relatively unaffected. This happens when all asteroids hit the planet simultaneously, during an extraordinary event.

This is hidden from you, intelligent life everywhere, and more importantly, life continuously everywhere, intelligent life. You might be determined to ignore and reject all these, yet there are people who remember these, as they use this type of information in order to control you in every manner.

Science asks questions as 'how did life originate on Earth,' or 'how did humans originate on Earth.' Then science lets the crowd fight throughout endless empirical debates while it sits back to relax, never providing an answer. It is only a scientific strategy, hidden in the question itself, since Life and humankind might have not even originated on Earth. There have been reptiles around, for hundreds of millions of years before mammals and humans, since the old environment allowed it, yet there were a multitude of other species on Earth before the reptiles. Earth is too cold for many species of reptiles, their niche is not here anymore, and therefore they cannot claim Earth anymore. It is stated in Bibles that the humankind has the right to roam the surface of Earth, and the snakes the underground. This is the case today, stated by Higher Laws.

Science also makes you believe that Earth has always been as it currently is, yet even humans were not always as they are today. Only twenty thousand years ago, elephants were very large, they were the mammoths. The lions and tigers hunting them were very large, and then even the people hunting mammoths were considerably larger than what they are today, they were the Giants, the inhabitants of Gaya, or Earth. Not only these, but all species were larger in size. Dragonflies were one meter long, which is impressive for a flying insect. Scientists claim that excessive oxygen enlarged animals. This could be the case, or the abundance of vegetation caused more oxygen in the air and more food in the niche of all species, allowing them to grow larger in all food chains.

It just happens that people on Earth are larger or smaller in one age or another, according to the entire environment, since the entire environment changes considerably from one age to another. People and life in general are much larger during or right before ice ages, and smaller during warm ages. Impressive giants roamed north-west Europe during the last ice age, wherever ice or absence of ice allowed it, with their large skeletons still being found everywhere, all validating the multitude of myths and legends from Europe and from everywhere else. Gilgamesh himself was a giant in size, yet only one quarter human, the way his myth states.

Yet this is not a rule. Throughout the warmer climate of Atlantis, people were smaller. They were probably smaller because they were not from those places, or because they were actually direct descendants of the previous civilization inhabiting the Earth during the past warm age.

These succeeding civilizations from within their own succeeding ages of Earth, warm and cold, have always fought each other for supremacy, for the supremacy of the same niche, with one thing defining their people, their size. Who won? Sometimes, specific remnants of the previous civilization managed to use their talent, knowledge, strength, and abilities, higher and lower, in order to subdue the entire newly emerging civilization. This is why you can find these individuals today portrayed as giants, very large in size, all being served by a multitude of shorter, smaller people, the crowd. The old deities of Egypt were larger in size, same as the old deities of Greece, same as the old deities of Mesopotamia, they were all larger, and more importantly, they were all the same. Yet if they were not exactly the same, then they were all part of the same family, the same genetic line, living their life casually, in the open, while interacting genetically only among themselves.

These are exactly the people or the genetic line still at play today, either in person or as direct descendants, or simply as impostors claiming genetic line, knowledge, wealth, and abilities today. From what we know, Alexander the Great found the last one of these remaining Sumerian deities, dead, in Bagdad, preserved in honey, at the end of his epic conquest to take over the world. Alexander the Great also claimed direct genetic lineage from these people, as many presidents, royalty, and famous politicians did.

Is this genetic line dead? Is there a difference between facts and beliefs anymore? Both the Brotherhood and the Elite claim that that entire genetic line has been resuscitated in every manner, and it is back in power, for centuries or millennia, rising once more to rule victoriously the way it has always done.

The people of Earth were larger when the word Giant appeared, and therefore this word is still in use today to define very large people. The word Giant itself means inhabitant of Geea, inhabitant of Earth. Only people smaller in size never refer to people larger in size as being giant. Etymologically, Giant means inhabitant of Earth. Therefore, the shorter people either had shrunk in this manner during the coming warm age, or they came from anywhere else. Myths, legends, religions, and stories mention Giants, along with renowned heroes hunting and killing them, as David. Why would the followers of religions who believed in deities from above kill people from Earth? Was it a war going on, or simply hunting, the basic hunting activity? No, it was about some people taking over the niche of other people. Even more, could this hunting still be going on today?

Who wins? Which side are you on? Everything depends on your developmental level, since if you are at your intelligent human developmental level or higher, you do not want to have anything to do with all these, you do not discriminate among people genetically, you accept them, and even more, you seek to connect with them and learn from them, as much as possible.

This two-sided genetic competition still goes on today, with one side or the other always winning and losing throughout time, indefinitely. This fight or competition manages to divide the entire social pyramid of power in two, vertically, into two distinct wings, engaging the entire world in the fight, directly or implicitly. Or it does not engage you at all, once you know these, and once you manage to turn your back and just live your life normally.

From higher perspectives, only this war or competition matters, with all causes being irrelevant. Therefore, this war will continue further throughout the future orders, indefinitely, since both competing genetic lines are already intermarried within the Elite and within the Brotherhood, and now it is only a matter of 'purifying' the common genetic line, on the behalf of one genetic side or another.

All these genetic lines have originated around that specific area where Sumerians lived or is said that they lived. The Sumerians have also started religions, along with an entire civilization that resembles to the one we know today, thousands of years ago. The Sumerians claim that they are the ones creating the humankind from scratch, or they had only modified it genetically, by adding their own genes. The Sumerians have also started many common religions, influencing many current beliefs that you know well. It is possible that the Sumerians are the actual deities as the myths state, or that they have borrowed these specific beliefs from other people, from other civilizations, and only passed them on to our current civilization. Either way, they claim to have been or to still be the actual deities themselves.

Consequently, three different genetic lines were found in that area, and can still be traced there and everywhere else. The first genetic line is that of the Sumerians themselves, claiming to be deities. The second genetic line is that of the old inhabitants of Earth, considered savages in those old times, and still being chased down in every manner along with all native genetic lines of the world, throughout the global genocide, and this includes natives from South America and North America, along with natives from Africa, Europe, Asia, and Japan. Furthermore, the third genetic line comprises the new humans that the Sumerians claim to have created or only modified genetically from the old inhabitants of Earth, genetic line that still seems to serve them even today, genetic line that is also found in the invisible kingdom.

This is how you have the well-known Masses, Brotherhood, and Elite today, distinguished genetically and through social status, while this is discrimination and segregation. This is how you have the same conflict going on today as it always did, taking place from the bottom of society up, and this is called genocide.

The Sumerians referred to the old, free human inhabitants of Earth as animals, or savage animals, while only the humans that they have created were considered to be human and to have human status. While they were not human themselves, but deities, or this is what they claimed. This was the case back then, while even then, you could not distinguish between these three specific human races, all being almost identical, and most importantly, all being compatible genetically and could intermarry. This is the case today, with the Elite claiming to be deities, considering the Brotherhood actual humans, while the inferior part of the Brotherhood along with the Masses are considered animals, domestic animals, or cattle. This is how the members of the Brotherhood refer to the Masses everywhere, as being irrelevant, animals, or cattle.

Can this really be the case? Can you actually be inferior or superior, wherever you are positioned in society? Everything depends on your status, and most importantly, everything depends on the status that you claim for yourself. In general, if you claim an animal status, then that is how others are allowed to consider you, as an animal, or as a domestic animal. If you claim higher status, than that is your status. This is always the case, since your development always matches your abilities, awareness, reasoning, and intelligence, and therefore that is what you are capable to claim. Doves, bees, and ants cannot claim intelligent status, and therefore anyone can claim a second level animal status for them. This is how anyone who does so can have rights over these specific individual birds, insects, or animals that they claim. This does not stop here, since the Brotherhood and the Elite make continuous claims over the Masses, just by referring to them as savages or animals. If these specific people do not respond, or if they lack awareness to understand what is going on, they become the property of those making these claims, just as it happened with the bees and ants from the above example.

This does not means that you are an animal, and belong to someone else as cattle, without even knowing it, because once you become aware of everything, once you claim third level status or intelligent human status, you are free, you are an intelligent human being, you have your status, and the Higher Laws are at your side to guarantee your intelligent human rights. Yet you do not have to claim anything, because as a human being, you already have your human status and human rights from birth. It is only when you claim otherwise, that your status changes to whatever you claim to be, cattle, servant, or slave, if you choose to do so, and this is what you become, through agreement and full consent.

This is why slaves have to make an oath of servitude to their owners before they are considered slaves, and the Higher Laws allow it. Our laws here on Earth do not allow slavery and exploitation of any kind, which is good. Yet currently, you wave your intelligent human rights and intelligent human status whenever you state that you are a consensual corporation written as your name in uppercase letters, as JOHN SMITH. Search throughout your documents and IDs, to find this corporation everywhere, waving your intelligent human status and rights. That corporation can be owned, bought, and sold, as all corporations, which also your case, every time you represent this corporation.

As a last example, if anyone claims that you have been created or modified genetically by them and therefore you belong to them, entirely or partially, along with your genetic line, entirely or partially, this does not have to be the case, even when all evidences are present. Because all living beings are as free as they claim, with no exceptions. Once you lack the awareness of your status and condition in the world, or once you lack the abilities to state it clearly, due to natural or artificial causes, once you cannot claim an accurate status to state your freedom, then you are considered by default part of the social chain. However, it is different in an actual human world, because while developed to the intelligent level, you are considered equal to everybody else, in a continuous human harmony. While only within decayed worlds, all undeveloped people rush up to form distinct social hierarchies where they seek to climb as high as possible in order to be able to enslave and exploit as many people below as possible.

If you were ever wondering why authorities never inform you of the truth about your origins, about extraterrestrial life, about other beings out there, about beings, intelligences, or entities stronger and more evolved than you, stronger and more evolved than humanity, or about your soul, or if you have a soul, or what a soul is and what is your relationship with your soul, this is all done on purpose, to limit your awareness and therefore your status and rights. Specific wider and higher knowledge is kept away from you on purpose in order to limit and restrain your awareness, and therefore to lower your status all the way down, in order for all those who know and remember the truth to claim you, your descendants, or your higher selves as property, as servants, or as slaves, for as long as you remain ignorant. You give these away, while taking drugs.

If you have doubts concerning the Sumerian genetic abilities, just study all fruit trees, or study the common wheat plant giving you your daily bread, along with many other crops, since these were created back then, genetically or from scratch. Many domestic animals had been created there and then. However, study other crops and domestic animals used by humans in other parts of the world, to find them similarly modified or created from scratch, as the apples, grapes, and wheat from Mesopotamia, and the corn and potatoes from South America. However, all crops and domestic animals form South America date farther back in the past, before Mesopotamia, matching the age of all ruins and relics found everywhere there on the ground. Corn is an obvious example of two plants spliced together to produce one of the best crops still in use by the humankind. That specific very old civilization from South America did not produce only the crops and domestic animals, but they have also produced and modified the soil itself for all plants to grow, soil that is still the richest in the world, with the Sumerians, Mesopotamians, and all their descendants still incapable of such an achievement. However, this entire world is made to believe that the Mesopotamians and implicitly the Sumerians that were their deities, were the ones to have created this entire world, from scratch, including humans, plants, and animals, which is not true.

Since the entire revision meant to erase and censor the past is very determined to state that Sumer and Mesopotamia were the very first civilizations on Earth, with the entire humanity savage, now we can understand what goes on. The Sumerians took over an entire civilized, highly developed planet in a controversial manner, now they claim it as their own, they have erased all native genetic lines then, according to Bibles and old records, and they are still erasing their traces today, through this entire, continuous genocide.

Yet we cannot discriminate among the people of Earth, since we are the people of Earth, and this is our world. Even more, just through your presence on Earth, it means that you have direct lineage to the old Sumerians and the old Atlantians alike, which is the case with almost everybody else by now. However, for all undeveloped people of the world, it is only a matter of who is more superior and therefore more entitled to receive a bigger piece of the pie, since this is always the case in a decayed world.

The End

This book series continues with the next book, "The Human Society." Here is a short synopsis:
Is the human society fair and fulfilling as you learn in school, or it is harmful, with corrupt politicians, financial cartels, and major conspiracies spanning the world as you always notice? Both are the case, yet the human society

improves gradually, despite of all corruption in the news and in the world. Only that the media tends to highlight politicians in order to capture the masses, or you never watch the news. Because it is always a show, even in the news, meant to keep the world content.

Is society actually corrupt and harmful? The human society has always been exploitive, yet people tend to interact in any manner in the world, more or less humane, while it is significant to distinguish your own influence exactly as it is.

Would you like to learn the truth about the human society? Then study yourself and those around very well, since society is the direct interconnectivity of all human beings, with you in it. Therefore, you are the one defining society directly, at least in everything regarding you, and this is the case with everybody else. While everybody is relatively good intentioned in society, since we are very similar through our human nature and through all natural living needs and meanings that we fulfill.

Because humanity is never divided into the good and the bad, since everybody is good, by having similar natural needs and meanings, and therefore similar fulfillments. Only that people can become more exploitive while fulfilling consensual needs, as these make them be whatever their superiors, jurisdictions, and ideologies desire, stepping in this manner outside the actual human nature, right into the consensus. While this is bad, since this is how humans become unhuman, with all dreadful consequences manifesting in the world. Because humans fulfill mostly consensual needs, as orders and duties coming from above, but not their own human needs, as everyone should. While this ends up defining the human society the most, changing it altogether into a consensual social machine meant for profit and exploitation, which should never be the case in a human world.

Because there are two human societies to consider, the natural intelligent human society that everyone seeks to have and maintain through their own living natural needs and feelings, and the consensual society actually instated in the world, regardless if you want it or not.

Throughout this book, we study how life and living human beings gather naturally to form classes of life and living societies meant to make life better, safer, meaningful, and therefore fulfilling. We also understand the current human society in its consensual structure and characteristics, in all aspects and from all perspectives. Furthermore, we identify and understand the various modes of society that life and the environment demand, as we recognize what is meaningful among orders, agendas, and conspiracies already implemented or threatening to take place in the world, who the main social actors are, and how determined they remain in everything that they do in the human society. This helps you understand your own meaning in life, in society, and in the world, while understanding how your meaning is enhanced or altered by your own behavior and interconnectivity in life and in the world. If you seek to uncover and understand the human society exactly as it is and as it should be, this book is for you.

ABOUT THE AUTHOR

Valentin Leonard Matcas, M.Ed., is a researcher, physicist, mathematician, educator, and an author of nonfiction and fiction books, including the entire "Human" book series. Valentin Leonard Matcas wrote the "Human" book series in the following order: "The Human Needs", "The Human Addictions," "The Hierarchy of Needs," "Stay in Shape, Lead a Healthy Life," "The Human Origins," "The Human Society," "The Human Conspiracy," "The Human Mind," "The Human Reality," "Astral Planes and Your Other Realities," "Life," "The Hierarchy of Intelligences," "The Human Intelligences," "The Human Thoughts," "Mental Models and Successful Ideas," "The Human Attitudes," "The Human Stereotypes," "The Human Ideology," "Modes of Life," "The Human Development," "Patterns of Development," "The Human Lifestyle," "Heal Yourself," "The Human Civilization," "The Human Religion and Spirituality," "The Human Rights," "Higher Laws," "Natural Laws of the Universe," "Existence," "The Human Condition", "Lifelines of Causality," "The Human Behavior," "Flat Earth," "The Human Environment," "The Human Meaning," "The Human Reasoning," "The Human Interconnectivity," "The Consensual Matrix," "The Matrix of Life," and "The Human Knowledge."
Valentin Leonard Matcas writes about terrestrial and alien civilizations, about life in the universe, the way it develops and intertwines across galaxies, about powerful beings as they control and reshape the universe, and about normal living human beings from Earth caught in this beautiful, wider, outstanding interconnectivity. Valentin Leonard Matcas creates a living, warmer universe in his books, teaming with life and vibrancy, on all levels of existence. Valentin Leonard Matcas also wrote "The Storyteller" book series, including "The Storyteller," "Starship Colonial," and "Unlimited," and "The Culling" book series, including "The Culling," "The Dream of the Dead," and "The Last Man on Earth."

Valentin Leonard Matcas

When he does not work on his books, Valentin Leonard Matcas enjoys researching, hiking, swimming, kayaking, skiing, snowboarding, biking, reading, listening to music, and playing strategy videogames. You may discover all his books, videos, and articles.

www.ingramcontent.com/pod-product-compliance
Lightning Source LLC
Chambersburg PA
CBHW020631220526
45464CB00001B/98